Recomposed

Kyle Devine is a professor in environmental studies and dean of graduate studies at the University of Winnipeg. He is the author of *Decomposed: The Political Ecology of Music*, an award-winning environmental history of the record industry.

Recomposed

Music, Climate, Crisis, Change

Kyle Devine

London • New York

First published by Verso 2026

The manufacturer's authorized representative in the EU for product safety (GPSR) is LOGOS EUROPE, 9 rue Nicolas Poussin, 17000, La Rochelle, France contact@logoseurope.eu

1 3 5 7 9 10 8 6 4 2

Verso
UK: 6 Meard Street, London W1F 0EG
US: 207 East 32nd Street, New York, NY 10016
versobooks.com

Verso is the imprint of New Left Books

ISBN-13: 978-1-80429-817-6
ISBN-13: 978-1-80429-818-3 (UK EBK)
ISBN-13: 978-1-80429-819-0 (US EBK)

British Library Cataloguing in Publication Data
A catalogue record for this book is available from the British Library

Library of Congress Cataloging-in-Publication Data

Names: Devine, Kyle author
Title: Recomposed : music, climate, crisis, change / Kyle Devine.
Description: London ; New York : Verso, 2026. | Includes bibliographical references and index.
Identifiers: LCCN 2026010616 (print) | LCCN 2026010617 (ebook) | ISBN 9781804298176 trade paperback | ISBN 9781804298190 ebook
Subjects: LCSH: Music trade—Environmental aspects | Sound recording industry—Environmental aspects | Concert tours—Environmental aspects | Music—Environmental aspects | Concerts—Environmental aspects | Climatic changes—Social aspects
Classification: LCC ML3916 .D507 2026 (print) | LCC ML3916 (ebook) | DDC 621.389/32—dc23/eng/20260324
LC record available at https://lccn.loc.gov/2026010616
LC ebook record available at https://lccn.loc.gov/2026010617

Typeset in Sabon by Biblichor Ltd, Scotland
Printed and bound by CPI Group (UK) Ltd, Croydon CR0 4YY

Contents

Preface

This book is about people in music who are reshaping what they do in light of the climate crisis.

I have written the book for the people it is about. Most of the time, I refer to all these individuals simply as musicians. This is the case whether they are performers, composers, fans, students, teachers, researchers, journalists, activists, entrepreneurs, nine-to-fivers, or whatever.

I have also written the book so that it might invite readers generally concerned about the climate crisis, but maybe not its drier subjects, further into climate politics. It is important to talk about IPCC reports, variable renewable energies, green quantitative easing, stranded asset management, evapotranspiration rates, Atlantic Meridional Overturning Circulation, El Niño–Southern Oscillation, and so on. My hope is that the story of how music connects to the climate crisis, while perhaps less consequential than heavy duty economic policy or hardcore climate science, will appeal to people in ways those things don't. Music is like that.

About what is ahead: The introduction develops the ideas and vocabularies that I have needed to make sense of the book's subject matter. I set myself the task of laying the groundwork as plainly and engagingly as I could. It is in the chapters themselves, though, where ideas and vocabularies turn into stories and people. I hope the book, including the introduction, is relatable for most anyone with an interest in music or the environment—regardless of how much they already know about climate issues.

The three main sections of the book are thematic. They present the most prominent axes around which the music world has gathered itself to, in its words, solve the climate crisis.

Part I covers technical solutions, mostly in the record industry. It raises some questions about where technological hopes and criticisms should be anchored. Part II covers institutional solutions, mainly in terms of carbon emissions. It raises some questions about the meaning of atmospheric accounting and accountability. Part III covers cultural solutions, looking primarily at how the arts and humanities are supposed to awaken sensibilities that link climate awareness with climate action. It raises some questions about how music culture might accomplish that. While the first two sections are largely about how the music world is reshaping its own circumstances, the third section opens onto the more general issue of how music, as music, contributes to social stability and social change. Music and culture have a role to play in the struggle for the planet. It's just not the role that most people think.

A few notes about the text are in order. Since I appreciate both the leanness of US spelling conventions and the logic of metric measurements, the two systems are here mixed. Currency figures are in the US dollar unless otherwise indicated. Quotations have mostly been incorporated into the book's house style and occasionally edited for clarity. That said, I have attempted to write in a way that avoids flooding the text with names and recitations from secondary sources.

I compensate by providing endnotes that are full of references I have relied on in writing the book.

Another thing worth underlining, especially in a book that takes social stratification as a central issue, is that authors have a responsibility to locate themselves in their stories. I understand knowledge as a state of being in the world, and I know the slice of the world I've studied has revealed itself to me at the intersection of my own accidental histories. Equally, though, it is important to insist that the reality of ideas cannot be reduced to biography. I did my best to reflect on these premises during my research, asking what it means to study climate crisis with the material advantages of Norway's oil economy—I was living and teaching in Norway while writing this book—and to examine emissions reductions while emitting carbon. Asking why most of the people I met in my research have similar faces and live in similar places, why certain patterns of climate thought and action nevertheless persist across boundaries of citizenship, geography, and culture, and why class consciousness in the social stratum that carries those patterns across those boundaries—the middle stratum, my stratum—is virtually a contradiction in terms. I reflect on these contradictions throughout these pages, and I take up such questions explicitly in the acknowledgments.

Introduction
The Great Recomposition

So I ate the plate.

Eating a plate made of compressed, steam-treated wheat bran is exactly as appetizing as it sounds. Imagine chewing an old Masonite clipboard until it turns to soggy breakfast cereal.

Yes, this thing is dictionary-definition edible. Ads feature the product being enthusiastically devoured by a goat, one of the animal kingdom's most famously unpicky eaters. My advice? Embrace the plate's other selling point—its patented disposability.

What counts here is obviously not gastronomy. What matters is the satisfying feeling of doing something so positively, utterly green. And it's not just the dinnerware.

At Norway's biggest music festival, concertgoers can wash down their biodegradable bran plate, and the organic snack on top of it, with a local brew in a reusable cup. Whatever is not consumed or reused is recycled at a rate of 98 percent.

To get to the festival grounds in central Oslo, attendees can walk, bike, ride a bus or tram, or take the metro. Only 2 percent of ticketholders drive, which matters because most concert emissions come from audience travel.

Musicians, meanwhile, can sign a green rider. Just as performers have hospitality needs and technical requirements written into their contracts, they now also insist on environmental provisions. Less meat. Lights out. Reduce, reuse, recycle. Everyone is reminded that the whole extravaganza is powered by water and wind. It has been this way since 2009.

No wonder the Øya Festival is known as a leader in sustainability and as one of the greenest music events on the planet. At the same time, Øya is just one example of a wider reality—a reality that is currently sweeping the world of music.

While there are precedents going back more than half a century, to antipollution folk songs and pro-solar benefit concerts, the turning point that defines this new reality started taking shape around the year 2000. It was then that Bonnie Raitt started putting biodiesel in her tour buses and Radiohead began inspecting their carbon footprint. By the early 2020s, acts like Coldplay, the Roots, Massive Attack, Billie Eilish, and many more (including the entire genre of K-pop) were attracting media attention for their commitments to sustainability—whether they were generating electricity with off-grid dance floors or cleaning toilets with ecofriendly chemicals. In ways both dazzling and routine, music today is going green.

It's more than concerts. And it's more than stars. People everywhere are contributing, in all kinds of ways. They are inventing records made of plants, building stereos that run on sunshine, dreaming of streaming services powered like hot springs and cooled inside mountains. They are making instruments from responsible resources that have been traded fairly. They are establishing nonprofit and investment initiatives geared toward environmental concerns. They are introducing carbon calculators, emissions audits, and climate credentials. They are forming organizations that size up (and draw down) the environmental impact of music on all levels. And they are generating a lot of ecological awareness: white papers, green road maps, gray literature, yellow pages, industry pacts, corporate reports, news columns, college courses, emergency declarations, sound installations, song contests, podcasts, playlists, and more.

Everywhere you look, the music world is changing along these lines—overhauling itself in response to the climate crisis. Call it the Great Recomposition.

Music's recomposition is not happening solely on the level of songs and styles. Of course, it is happening there too. Music has always registered and distributed environmental sensibilities. This is why there are symphonies of scenery, psalms about paving paradise, concerts for causes, elegies on icebergs, dances of discobedience. But the current transformation does not depend on music's role as a barometer of the ecological imagination or as a catalyst of it. What is changing are the basic technical, institutional, and cultural conditions that determine how music gets made and heard in the first place. This is a climate-oriented transformation of what music is and how it comes to be, not just what it is about or how it sounds.[1]

The Great Recomposition probably seems like a good thing.

It is. Music has taken on the commendable task of large-scale climate action—of reforming itself, top to bottom, in connection with the climate crisis. The job is being done by principled, hardworking people who rightly believe in what they are doing. Their work is ameliorating music's environmental impact. It may even be helping to ameliorate the climate crisis more generally.

But it's not all good news. A variety of factors—bigger than any individual, product, or industry—threaten to get in the way of what everyone wants to accomplish. One main roadblock is that music's technical, institutional, and cultural reforms are thought of as "solutions" to a "problem" called "climate crisis."

Far from denying the reality of the changing climate, I am convinced that it is only possible to grasp the severity of the situation by learning to redefine the problem. It is necessary to think a bit less in terms of climate and crisis, a bit more in terms of the economic arrangement (capital) and corresponding social architecture (class) that define the world today.

The climate crisis is an effect of those more basic conditions, both concretely and conceptually. Projects that accept climate crisis as the problem, and that attempt to solve it at the level of appearances rather than essences, are currently required. But they are also confined to a form of solutionism that keeps the secrets it pretends to tell. From this perspective, attempting to "solve the climate crisis" can be understood as a euphemism that conceals a contradiction, consecrates an illusion, and curtains a reality.[2]

Climate solutionism masks the contradiction that inheres in pursuits of sustainability that do not question certain underlying economic forces. It sanctifies the illusion that fixing this warming world is simply an engineering challenge, a market failure, or a knowledge gap with correspondingly benign technical, institutional, and cultural remedies. It disregards the reality of the planetary factory that will be required do the heavy lifting in any transition to a more environmentally friendly economy.

For these reasons, climate solutionism tends to turn even the best ideas and intentions back into the effects they set out to oppose. But this is no standoff. Real change will come from both the recomposition and the recomposition of the recomposition. That's the message here. It is the old acting strategy of *yes, and . . .*

Our story begins where another leaves off.[3]

Previously, I wrote a book called *Decomposed*. That book is about how the record industry has taken advantage of planet and people alike and how the situation has gotten worse over time, even in the seemingly immaterial era of digital streaming. In fact, the whole music world has long been caught in a disturbing cobweb of exploited resources, workers, and much else. *Recomposed* takes all this as given. It asks, instead, What do we do?

At first, this was not my question. I certainly didn't plan to write a book about it. In fact, I did my best to dodge the issue—in *Decomposed* itself and in the coverage that followed. Of course, there were plenty of readers who dismissed the book as wrong (*No way streaming emits more than vinyl!*) and wrongheaded (*Focus on the real polluters!*). Those are reasonable reactions. I will return to them.

Most readers, though, seemed primed to accept the premises and arguments of *Decomposed*. Maybe they were surprised by the environmental cost of music. But they did not need 300 pages describing the mess. They wanted to know how to clean it up.

Some of those readers thought I might tell them how to do that, and they contacted me to find out.

I always knew that I didn't have the answers. Pretty soon, I figured out that I didn't have to have the answers. The work of recomposition was underway. Eventually, though, I got curious about the question.

What do we do? Having spent years talking to hundreds of musicians from around the world, I find that the question tends to have two meanings. First, it refers to climate adaptation:

How do we make music more sustainable? Second, it refers to climate mitigation: What can music do to help solve the wider climate crisis?

The former question is the main concern of this introduction and the first two-thirds of the book. The latter question comes into sharper focus as the introduction develops. It is the primary concern of the book's third section. My answers to both of these overlapping questions—answers that try to understand something about how the questions themselves have been formed—are based on the ideas and vocabularies developed in this opening chapter.

In asking how to make music more sustainable, people are usually founding their curiosity on one of two cardinal visions of sustainability. Some want to figure out how to sustain music culture as it currently exists. This is an industrialized version of music production and consumption that has taken shape over the past 150 years. It is a historical formation that expects and delivers a kind of 24 hours a day, 7 days a week, 365 days a year, all-you-can-eat buffet of music at bargain basement prices. The assumption here is that we should continue down this path. The expectation is that we can do so indefinitely—as long as we find the right technical, institutional, and cultural solutions.

Other people begin from the assumption that music has gone too far down the paths of industrialism and consumerism. A sustainable future for music, from this perspective, means turning back. It means we should expect, or practically beg for, restrictions and reductions. We should attend fewer concerts. We should stop buying so many records. We should go on a digital diet. We should want less music.

In their own ways, both answers are presented as progressive. In their own ways, as we will see, each is deeply conservative. For now, the point is to highlight something else. There is a reason these two opposing responses and their visions of sustainability offer themselves so readily and appeal so naturally to our climate dispositions. The question these responses attempt to answer already takes for granted the problem it wants to solve.

But climate crisis is not the only way of understanding the mess we face. It is a specific problem. And if the definition of a problem determines the scope of its possible solutions, what other problems are possible and how might they be constructed?

The first step here may be counterintuitive, because *climate crisis* has recently been adopted as a term with more urgency than earlier ones like *climate change* and *global warming*. But the term has quickly come to carry an assumption of consensus about its meaning, and despite sounding critical it actually forecloses possibilities for political dialogue, contestation, and disagreement. Although *climate crisis* does appear throughout these pages, because it is the operative term in the settings this book is about, it is not my term. Its constituent elements—*climate* and *crisis*—are worth taking apart to see how they work and what work they do.

The term *climate* calls attention to greenhouse gases and their atmospheric effects. This is the image that animates mainstream environmental thinking and underpins the climate common sense that dominates the news and most informs the music world. It is an image that glosses over the more general ecological disaster unfolding across the planet and marked by ocean acidification, land aridification, aquifer depletion, species extinction, and more. The emphasis on climate in mainstream environmental thinking sidesteps most of this, in favor of a simpler vision of atmospheric

Protester at COP26 in Glasgow, 2021. Photo by Matt Brennan.

accounting in which the books are forever cooked, concealing its real expenditures, debts, and effects.[4]

More and more people today recognize that our particular economic system, with its particular thirst for profit, is the source of the mess in front of us. Headlines blare across newspapers like the *Guardian* and the *New York Times*—"Capitalism Is Killing the Planet"; "The Next Reckoning: Capitalism and Climate Change"—while similar views are expressed throughout the social sphere, from tweets to treatises and every podcast in between. The sentiment was perfectly captured by a protester at the United Nations Climate Conference in Glasgow, also known as COP26. "It's capitalism, ya eejits."

This is true, but only partly. There are reasons to believe that capital could be capable of solving a problem called climate crisis, in the specific sense of atmospheric decarbonization. This is because climate crisis, seen largely as a matter of emissions, is partly a problem capital has set for itself. What capital cannot solve, though, and what gets disguised by portraying the climate issue primarily in terms of carbon dioxide emissions, is twofold.[5]

First, capital's internal logic ensures that, even if carbon emissions are drastically reduced or eliminated in production and consumption, the pressures of competition and profit will continue clobbering nature in other ways. The effects of this are sure to pop up elsewhere, although the results of such planetary whack-a-mole are not predictable. They will take time to emerge. The devastation of nature is one of capital's main enduring features. It is a feature that may be diffused or displaced, perhaps even very creatively so, but it probably cannot be overcome. Whatever limits there are to capital, they are inside it—not outside.[6]

The second insoluble difficulty is capital's specific tendency toward economic growth or the particular way it creates new economic value. At bottom, capitalist economies are capable of this achievement because they withdraw uncompensated value from the paid and unpaid work people do. Capital also tends to externalize the full human and environmental burdens of making goods and providing services. These dimensions of value creation are foisted onto badly treated populations and places. "Growth"—not economic growth as such but capitalist growth in particular—is a polite word for exploitation and oppression.

The *Guardian*, the *New York Times*, the protesters—they're all right. Partly. Capital is to blame for the altered climate. But simply calling out capital is not enough. In fact, anticapitalism is good business. Confronting capital without equally confronting the social order that it makes and remakes will not upset the arrangement that, to be put right, would have to be abandoned.

For these reasons, the focus on the climate that dominates mainstream environmental thinking can perpetuate parochialism—often unintentionally so, despite its intentions. Take, for example, Coldplay's widely publicized 2022 world tour. Much of the band's promotion, and much of the media's coverage, focused on emissions targets. (They pledged a 50 percent drop and managed to hit 47 percent.) Among numerous reduction measures and sustainability initiatives, there was no shortage of hype surrounding the role of custom-built, electricity-generating kinetic dance floors in lowering the tour's on-grid electricity usage. The off-grid dance floors played a significant role in placing Coldplay "among the most environmentally friendly major touring acts in the world."[7]

What no coverage mentioned, though, is that things that appear environmentally friendly to some, like the band members and audience members of a low-carbon Coldplay concert in, say, Amsterdam, may be experienced differently by others, like those in the Netherlands who build the band's kinetic dance floors. Or those in the Andes who mine the lithium for those floors. Or those who profit most from the lithium mines and can afford to jet to Amsterdam and dance to Coldplay.

This is one way that emissions-first climate common sense falsely universalizes solutions that make perfect sense for some people in some places, with particular understandings and experiences of the problem, but make a lot less sense for others. Climate solutionism preserves unequal and unjust relationships between the economic core, or the North, and the economic periphery, or the South, as well as between upper and lower social ranks in the core and periphery themselves.

Like *climate*, *crisis* does specific work in troubled times. Although the term's meaning seems straightforward, crisis designations are never simply descriptions of the world. They are active moments in labeling and shaping our reality.

In the same way that the possibility of a copy creates the idea of an original, the designation of a crisis works to delineate what is normal. Yet crisis designations call attention to themselves, to the state of emergency, rather than the conditions that give rise to crises. Improbably, the result is that crisis itself becomes normalized. In an economic arrangement defined by capital, seeking to understand the reproduction of that system and seeking to understand its crises are the same task.[8]

If crisis is an operation of capital that today directs attention toward the climate, then it is possible to see climate crisis as an operational protocol that, far from disturbing capital, actually serves to reenergize and secure it. Researchers have long understood that crisis designations betray the literal meaning of *crisis* as a turning point. Economically, crisis is more like a through road.

From this perspective, although it speaks a language of emergency and upheaval and change, crisis conceals stasis. What music's recomposition bears witness to, as with the wider energy transition that it mirrors in microcosm, is a transformation of what some specialists call the regime of accumulation. It is a shift from one way the economy reproduces itself, based on fossil fuels, to a different one, based on decarbonized energy. This is an upshift—not a downshift—from fossil capital to green capital, and climate crisis is the clutch.[9]

Solutions that offer themselves to the problem of climate crisis are providing real relief. But these solutions cannot escape certain restrictions. The most imprisoning restriction is that capital is fully capable of profiting from the tremulous planetary situation that it created, then denied, then openly acknowledged, and now proposes to solve—and make a buck on—and therefore to perpetuate.

So it goes with music and much else these days. Yet the purpose of this book is not only to describe the situation. It is also to explain the situation's persistence and to suggest how to change it.

The question of persistence can be answered by looking at certain differences between economic priorities and ecological principles in capital's social architecture. As an economic arrangement, capital brings this architecture into being and also depends on it for its reproduction. In the process, it sets people into different groups, or classes, and choreographs them around competing interests.

Interests are simply what is beneficial and advantageous to individuals as well as to collectives with common concerns.

The complication with class interests is that they are often pulled in two directions at once. For example, it would be in the immediate interest of the event staff at our Coldplay concert to get together and insist on better wages. But pushing for wages alone would not address the more fundamental interests of the employees. Addressing those interests would involve questioning the larger economy that is based not only on their relatively low pay and lack of say in their working conditions but also their absolute lack of ownership. In an unjust twist, pursuing one set of interests usually comes at the expense of the other.

Mainstream climate politics are like this. They tend to be vitalized by a particular set of competing demands that define a contradictory place in our social architecture known as the middle class. The whole idea of a middle class is controversial for a number of reasons, but it does provide a specifically valuable way of explaining some prevailing patterns of climate common sense.[10]

The middle class can be seen as a social space where people have both the sensibilities and the means to care about the world in particular ways. These sensibilities and means find expression primarily in the realms of the individual and the entrepreneurial, which conform to the classically middle-class historical figures of the consumer activist and the conscious capitalist, or the sympathetic consumer and the sympathetic producer.

Contemporary sympathies may come to mind, like the ongoing calls to boycott sweatshop sneakers and Foxconn phones. Or the impulse to support (and found) brands that feature fair trade farmers smiling on bags of chips and coffee, and the pull to patronize (and establish) restaurants with a story of the supply chain behind every dish, the microclimate in every bottle. From the opposite affective register, there is that tinge of guilt at forgetting a reusable grocery bag. That moment of hesitation before declining to pay the emissions offset fee that promises to neutralize your impact on the climate.

The examples are everywhere. But the archetypes of sympathetic consumption and production are as old as capital itself. Take, for instance, the widespread consumer movements around the turn of the twentieth century, such as the National Consumers League in the United States, which drew attention to the laborious "Christmas cruelties" that each season underpinned the joy of giving. Further back, around the turn of the nineteenth century, there were the British abolitionist movements whose pamphleteers informed consumers that commodities like rum and sugar were "polluted, physically and spiritually" and that good Christians had better take their money elsewhere.[11]

Whether taking shape around 1800, 1900, or today, such sympathies exhibit remarkable consistencies in outlook and political vision. They all think of the world as a consumer-centered network of production, circulation, and exchange. They have all "understood consumption in political terms and have practiced politics in consumerist terms." And they all lack a sense of their own past, with minimal knowledge "that others before them had fought analogous battles, used similar tactics, and shared an understanding of the ways in which markets linked distant people." At no point in this history have sympathetic consumption or production been very effective in easing the suffering and damage that capital brings.[12]

Today both the sympathetic consumer and the sympathetic producer want to do right by the climate. They are appalled by capital or at least sensitive to its excesses. They aim to resist and alter the arrangement from within or at least to find an ethical relationship to it. In music, this means passionately debating topics like which format is the most environmentally friendly (vinyl or streaming?), which mode of touring emits the least carbon (bus or train?), which songs spread the most awareness (anthems for the Anthropocene or playlists for the planet?). Yet such sympathetic forms of production and consumption are generally not inclined to confront capital or class themselves. While doing so would address certain fundamental interests in human freedoms and planetary futures, taking that step would come into conflict with more immediate interests like satisfying basic needs, such as earning money for sustenance and shelter. Contrary to the many possible solutions to climate crisis, solutions to the exploitation and injustices that define the problem of capital are unlikely to be profitable and unlikely to be accommodated.

Imagine that the employees who make Coldplay's off-grid dance floors invent a miraculously efficient version of the product, one that promises to power not just some small fraction of a Coldplay concert, as the current ones do, but the whole event. But imagine too that the parts and labor required to build the new dance floors are so expensive that they are not obviously affordable or profitable. Perhaps the new technology would raise ticket costs to a point where fans would no longer be willing to pay the price. The result would be the same. Employers—bosses and managers acting not out of malice but simply in their own best interests—will give the green solution a red light. The employees may consider going out on their own, starting their own virtuous company, but that would be a big risk. Their current pay is steady, and they have people at home who rely on them, so they resign themselves to their current situation. In such cases, immediate economic priorities are, very reasonably, put before fundamental ecological principles—and the existing dance floors persist, because they do some good and turn a profit.

The priority of immediate interests is not secured because our dance floor engineers (sympathetic producers) or our Coldplay fans (sympathetic consumers) are under the threat of violence or the spell of brainwashing. Rather, it is secured by the dull thrum of economic relations—by economic power. Pursuing ecological principles (which is in everyone's fundamental interest) only coincides for so long with the pursuit of economic priorities (which is in everyone's immediate interest). At a certain point, when presented with an opportunity to confront capital head-on, principle and priority typically reach a compromise, while resistance and resignation fold into a knowing embrace. This is one humdrum way that economic power works to limit its own confrontation.[13]

Nobody I spoke to in my research was fooled by any of this. The music world understands what it is up against. Nevertheless, ideology does have an explanatory role to play here—as something less crass than false consciousness but more potent than plain old socially conditioned thought. Ideology helps to adjudicate the real tensions that animate middle-class climate politics. It works as interpersonal scaffolding where a particular range of actions offer themselves as more and less reasonable, more or less achievable, than others. We can refer to this coordination of the material and the ideological in terms of ecoliberalism—a phenomenon that is every bit as prominent on the left of the political spectrum as it is on the right. In fact, much of this book can be understood as a sympathetic left investigation of a pervasive and broadly leftist middle-class ecoliberalism.[14]

In explaining the persistence of climate solutionism by pointing to middle-class materialism and ecoliberal ideology, I am not claiming to know anything about the individuals involved in the work of recomposition—not how much money they make, how many degrees they have, where they grew up, which way they vote, what their parents do, or anything like that. I never asked. I rarely was told.

It would certainly be possible to guess. But my concern is not with the predictive power of class in relation to demography. My concern is with the restrictive power of class in relation to possibility. Class in this book is understood positionally, not personally.

Although I have described music's recomposition as a middle-class and ecoliberal phenomenon, one that contributes to the persistence of capital more than disrupts it, and therefore has a tendency to become the problem it wants to solve, this is not to foreclose the class's political potential. Nor is it to foreclose the political potential of music. Both may contribute to change.

Maybe it seems strange to think about the middle class in such terms. After all, there is plenty of evidence that the middle class serves conservative and reactionary political agendas. Arguments here range from the richly historical (governments and intellectuals have collaborated with capital to neutralize the middle class by freeing it "from its historical relationship with work in an attempt to emancipate it from the field of tension into which it was continually dragged by class conflict") to the entertainingly polemical (the middle class is capital's "most assiduous courtier and sycophant")—and everything in between.[15]

If the political potential of this class has been dismissed, even demonized, there is also reason to believe that the middle class—the definition and invocation of which always express motive—has progressive potential. In fact, the class's radical acts and alliances are matters of historical record. This returns us to an age-old conundrum about the role of the middle class in social stability and social change, about how such a contradictory location in capitalist social relations might be brought into confrontation with capital itself.[16]

The most effective way of confronting capital is to put pressure on its main vulnerability. By stopping the basic work of making goods and providing services, the top floor of capital's social architecture is made to listen, because the economic arrangement cannot function without its ground-floor workforce. This is the leverage of labor. The crucial question, then, is how the middle class might become a class, neither in itself nor for itself, that helps pursue the fundamental interests of everyone. Alongside book clubs, movie nights, baseball leagues, blood drives, picnics, social media, and much else, music will play a role in organizing those political cultures of coalition. As it always has.[17]

Our economic arrangement has many pressure points, not just labor. Our social architecture has many forms of inequality, not just class. And political ignition may come from anywhere—from environmentalism, certainly, but also feminism, antiracism, anti-imperialism, antimilitarism, and so on. There is a lot of work to do on all those fronts, in terms of their specific causes—all of which are part of the complicated story of climate crisis, several of which feature in this book's story about musical recomposition.

Yet complexity in itself is not always the most important finding of empirical or theoretical work. It does take serious effort to identify the complexities in any situation, most of which are not readily observable on the surface of things. Having done that work, it then takes an equally serious effort to boil down the complexity of a given situation, to suggest an order of contributing factors, and to further reduce things to an explanatory simplification. In these pages, that explanatory simplification can be summarized as follows: The road to solving the climate crisis may well bypass a confrontation with capital. But the road to a livable and egalitarian planetary future—for everyone, everything, everywhere—goes through a class-based confrontation with capital, not around it.[18]

This is why, when students ask their environmental studies professors how to save the planet, the best ones don't just talk about reusable bags or carbon offsets or Coldplay concerts. They tell their students to join a union.[19]

It is also why, from now on, if people ask me how to make music sustainable or how music can help solve the climate crisis, I finally have an answer—and not just a question about the question. Keep going with all of music's technical, institutional, and cultural reforms. But join a union choir too. Or simply talk about this stuff at work, with neighbors, and in communities. Get involved in any socially progressive, politically active, structurally organized, ground-up radical mass movement that considers work and class. And maybe, if music is your thing, lead a chant at the next strike, dance a step at the next protest, bang a drum at the next demonstration, or sing a song at the next action.

It is easy to brush off radical questions as so much juvenile pie in the sky, to assert reform as the only realistic and grownup way forward. Reforms are needed. History is full of them—good ones—and so is this book. But radical questions need to be asked and answered too. Otherwise we risk unintentionally sticking to solutionist ways of thinking and being that prioritize the welfare of some (us, ourselves) at the expense of others (it, them). This is why radical authors refer to the seductive realism of reform as an "ideological counterinsurgency" that works "against those who not merely hope for, but need, a better world."[20]

Yes, and welcoming a change in the sea prevents no one from also appreciating a breeze across a lake.

August 8 was a beautiful day. People feasted and sang. They danced and got drunk. Cymbals crashed in the warm summer air as the sun and moon traded places in the sky. Mitski, Sigrid, Erykah Badu, and many other musicians—all played remarkable sets. What does it say, then, that my abiding memory of the 2019 Øya Festival is that damn plate?

It tells you something specific about me, for sure. But given the consistency with which festival promotions and reviews also mentioned the edible eco-plate, my experience may also speak to something more general about our moment in music history.

I have written *Recomposed* for that moment—a moment when it seems like everyone in music is wrestling with their world's environmental cost and restlessly seeking its potential. I've also written this book out of a belief that people should know what lies beneath the bright green surface of climate solutionism. They should know a deeper viridescence.

The point is never to be dispiriting about the Great Recomposition. It is an appeal to see it for what it is, true to life, and an invitation to realize what it might become.

Part I
Technical Solutions

1
The Ecological Record

"Slice my veins," the CEO says, teasing the audience in front of her. "Little plastic pellets will probably come out instead of blood."

Liz Dunster makes her living in plastic. She is the founder and CEO of Erika Records, which has been pressing vinyl in factories near Los Angeles since 1981. Dunster was the first and, for a long time, the only woman to own such a plant.

Erika makes all kinds of records. This includes many standard black LPs for big name acts. But Erika is equally known for its small-run anomalies: custom shapes, picture discs, clear vinyl implanted with "love notes, yarn, peacock feathers—almost everything the FDA will allow." One implant the Food and Drug Administration will not allow is maggots. The agency also ensures that, even though Dunster bleeds plastic figuratively, her records may not bleed blood literally.[1]

Dunster is fond of the vinyl-in-the-veins imagery. She uses it often, including at Making Vinyl, an international business meetup where I joined the Erika Records founder in Hollywood as part of a panel in 2019 called "Vinyl: More Sustainable Than You Think." According to the event program, "today's record industry is not like yesteryear's from the perspective of raw materials." These days, technical improvements have made things much greener, and to underline this point we were joined on the Making Vinyl panel by other "environmental-minded" representatives from a plastics supplier, a printing company, and a record label.

Spinning vinyl like this is to be expected in a business setting, but maybe not for the reasons you would think. It would be easy to dismiss the Making Vinyl panel and initiatives like it as mere marketing tactics or hollow social responsibility initiatives—attempts to make an industry, its companies, and their products seem environmentally friendly when, under the surface, it's business as usual. This is greenwashing, a term that conjures images of cynics and villains in corporate boardrooms.

But the Making Vinyl community is made up of good people doing good work. The same is true of most everyone pursing technical solutions to climate crisis, in music and beyond. For these people, greenwashing is not some evil plot. It isn't really a choice. As companies compete for advantage in market spaces increasingly converted to the doctrine of green capital—a theology in which capitalism has given us the climate crisis yet capitalism will also deliver us from it—putting a green spin on things is simply good business.

Green spin also reflects a structural imperative that limits the potential of technical solutions to climate crisis. What's good for profit and what's good for planet tend only to coincide for so long. There are disincentives to pursuing technical solutions past the point of no returns. This is an effect of a social position in our economic arrangement where businesspeople (as well as consumers) have their ecological principles set in tension with their economic priorities.

In cases where music industry personnel bring ecofriendly ideas to the table, for example, the instantaneous question from the higher-ups is always "What does it cost?" Bosses are not morally opposed to such ideas. They are just doing their job. It does not matter how green a solution is if it threatens to put a company in the red. Even the most potentially game-changing technical innovation is unreasonable to pursue if it will sink a business. That is an obstacle built into our economic arrangement, not a fault in anyone's moral fiber.

All this helps to emphasize that most people working on technical solutions to climate crisis are acting in good faith. They believe in what they are doing, and they should. Technical solutions will be part of any serious effort to address climate crisis. However, so long as such efforts take for granted their underlying business model, which is driven by specific tendencies of profit and reinvestment, technical solutions are bound by limitations. For those who want to work to the best of their ability within such constraints, and for those who wish to move beyond them, the limitations are worth understanding.

"Vinyl: More Sustainable Than You Think?" I always thought it should have been posed like this—a question rather than a declaration. Then again, debate was not the mandate for Making Vinyl's inaugural panel on sustainability. The point was to highlight environmental advances in the record industry. So that's what the panelists did.

For her part, Liz Dunster talked about the possibility of pressing discs using recovered ocean plastics. The Erika Records CEO also emphasized that all her vinyl is made in the United States, relaying her experiences with suspect supplies from other countries. Dunster additionally noted that Erika had been making unleaded records since 2015, a move

Making Vinyl in Hollywood, 2019.

she puts down partly to being a parent and grandparent. "I'd hate to have a record that had lead," she says, "and then a child starts gnawing on it."[2]

Ashby Baum of Westlake Chemical, a Houston-based Fortune 500 petro-polymer giant, also distanced his company from the poisonous heavy metal. Since LPs are made from a type of plastic that breaks down when it gets hot, as records do during the pressing process, they need a stabilizing additive. Lead—dense but yielding, silvery but ashen, handy but deadly—was long used for this purpose. Although some attendees were unsure what all the lead-free fuss was about, given that European suppliers had been using safer stabilizers since 1998, Baum reported the news as a minor triumph. Westlake was transitioning away from lead.

Baum also noted that his company used a significant amount of regrind. This is plastic that has undergone at least one round of processing, been rejected for some reason, then transformed back into raw material and reused. Record enthusiasts debate the role of regrind in the performance of their plastic. Some say it is beneficial. Others, detrimental. Westlake's position is clear. Records sound best in the sweet spot where cost saving and planet saving coincide.

Moving from discs to their packaging, Sam Gay was on hand to represent Stoughton Printing. Stoughton has been making record labels, jackets, and inserts in the City of Industry, California, since 1964. Gay stressed that the company uses recycled paper and soy-based ink. He also noted that the factory had gone down to a four-day workweek, partly to eliminate commuter emissions.

Stoughton's website, meanwhile, does more than advertise the company's current greenness. It says Stoughton was operating an "environmentally responsible manufacturing facility" before it was "hip" to do so. As early as 1985, ahead of ever-tightening California controls on air and water pollution, the company transitioned to an aqueous printing process that is less toxic than solvent-based methods. And sometime before the Forest Stewardship Council started certifying responsible woodland harvesting in 1993, company founder Jack Stoughton apparently had the environmental foresight to source paper from mills that farmed their own trees. Notwithstanding the contested nature of silviculture, they are doing good work.

Then there was Parks Vincent from Ninja Tune, an independent record label. The story of Ninja Tune's sustainability, as Vincent told it, owes something to a musician on their roster. Jayda Guy—Jayda G in the music industry—is a DJ and producer and also an environmental scientist. Around the time Jayda G released her 2019 album, and partly due to her insistence, Ninja Tune started offsetting their emissions by planting trees. They did away with single-use plastic packaging like shrink-wrap. They were working to cut down on employee flights as well as product shipping. By 2021, among many other climate initiatives, they had cofounded Music Declares Emergency and the Association of Independent Music's Climate Action Group. Ninja Tune was also a founding investor in a carbon calculator tailored to music's independent industries, developed by the IMPALA Independent Music Companies Association and the nonprofit arts-and-environment organization Julie's Bicycle. The label additionally announced plans to go carbon neutral by year's end and carbon negative thereafter.[3]

Ninja Tune's activities open onto a broad set of developments in the music world's institutional and cultural responses to climate crisis: awareness campaigns, emergency declarations, carbon calculators, offset initiatives, and more. Here it is worth reflecting on the rhetorical work of greening vinyl, and music more generally, because it illustrates something about the odd fellowship of economy and ecology.

It was not hard to see what was happening on the Making Vinyl panel. It benefits businesses to present themselves in the best light—which, these days, is a progressive green light. In the same way that a multinational energy company like British Petroleum can change its logo to a sunflower without addressing its dependence on fossil fuels, or how a global fast-food chain like McDonald's can dye its brand emerald without addressing its reliance on harsh factory farming, so can the vinyl record industry present itself as ecologically sound without addressing the fact that its product is fundamentally made of oil. This elephant was in the room at Making Vinyl. For all their environmental initiatives, none of the industry representatives on the panel was asking what the LP looked like after oil.

Greenwashing is funny that way. At worst, it is deception. But the conference speakers were simply emphasizing all the positive differences that are possible without addressing certain underlying issues. Appearance-level technical solutions are stuck between this rock and that hard place. Green spin is the unavoidable result of good-faith people inhabiting a bad-faith system.

As for the panel's nonindustry representative, described as a professor who proved "digital music leaves a bigger carbon footprint than physical media," that was me, but that is not how I would describe myself.

For one thing, the contrast between digital music and physical media is misleading. Digital music is physical music—because digital media are physical media. This is established, even if the facts are slow to influence how most people talk about online activities. All digital scenarios are highly coordinated physical situations. They rely on infrastructures of electricity and telecommunication, facilities of data storage and processing, devices of all kinds, and much else. But this is my ax to grind. What matters here is the panel description's implication that, from a carbon perspective, streaming is worse than vinyl.

It is true that, based on some estimates, the recording industry's carbon emissions appear higher with streaming than they were at the sales peaks of previous formats such as the CD, the cassette, and the LP. There are many ways to calculate the figures—and just as many ways to debate them.

For a single person to listen to a single album, there is no comparison between streaming and vinyl. One-to-one, streaming is vastly more efficient—vastly less energy-intensive—than producing earlier "physical" formats. From this perspective, streaming has a much smaller carbon footprint. Streaming an hour of pop music today emits about the same as microwaving six bags of popcorn. Five hours of streaming, though, can send up more emissions than a "physical" album. Yet one-to-one comparisons miss the point, because they confuse technical evaluation with social evaluation.[4]

The misunderstanding comes from calibrating the environmental consideration of digital music solely against the technologies of digital music. Even the miraculous efficiencies of computers, code, and compression are no match for the economic system that puts those technologies to work. That system is what gives music streaming its substantial emissions profile.

Tens of thousands of songs and albums, as well as, apparently, a gazillion podcasts, are added to streaming services each day. As if that were not enough, all this content is stored not just in one place but many. It is duplicated thousands of times across thousands of digital service providers that are locked in competition. This is a major inefficiency of the current worldwide internet infrastructure.

To take just one example, South Korea's biggest subscription streaming service, Melon, provided 3.7 billion hours of listening in 2021—thereby emitting an estimated 200 million kilograms of carbon equivalents. Moving forward, it is not just conventional streaming that will contribute greenhouse gases in recording's industries. Universal Music forecasts billions of dollars of digital industry growth due to smart homes, smart cars, and TikTok, while *Music Business Worldwide* observes: "virtual concerts, derivative works, the rise of companies selling sound packs and beats, AI music, etc. All of it signals a coming revolution that's going to transform the way music is created and experienced." Whatever music's next technical revolution may be, the industry's commitment to capital is not going anywhere. Neither is its emissions profile.[5]

This is where the organizers of Making Vinyl are right. Streaming has a bigger carbon footprint than vinyl. But this might not mean exactly what vinyl advocates want it to mean. The reason has little to do with streaming itself. It is because a large industry has figured out ways of selling more music to more people, more of the time. Were the music world to somehow graft this digital economic and cultural formation back onto LPs—were the vinyl industry's dreams to come true and LPs were suddenly to become dominant once again—it would be much harder on the climate.

A broad, parallel example brings the point home. The power of technology, all else being equal, ought to make things more efficient. Machines, for instance, should enable people to work less—to shorten the working day. Imagine it takes one worker eight hours to sew ten garments, while a machine can do twenty garments, more consistently, in four hours. This should be good news for stitchers. They should be able to set up the machine, produce twice as much in half the time, and spend more quality moments enjoying life. Yet when machines are used in the service of unchecked capital, the working day actually tends to get longer and more demanding. What's more, the financial benefits of increased productivity tend not to end up in the wallets of the people who run the machines.

Famously, this set of problems was noticed by the Luddites of the 1800s. The Luddites are condescendingly remembered as backwards-thinking technophobes because they took out some of their frustration with machine-based textile manufacturing on mechanical looms themselves. Yet what their movement actually protested was the incursion of capital into work and life. Similarly, although music streaming may generate more emissions than previous formats, this is because of how music today works as an industry—not because of how streaming works as a technology. It is a good reminder of the questionable wisdom of directing both protests and celebrations of technology primarily at technologies themselves.

If none of the Making Vinyl panel's industry representatives was thinking beyond oil, it became clear afterward that other people were. Someone in Russia was supposed to be making LPs from potatoes. True or not, the tuber rumor takes its place in a fanciful (sometimes actual) history of making records out of foodstuffs ranging from chocolate to coffee to cheese. There was even a Norwegian psychedelic supergroup that once hoped to release a record made of licorice—or, depending on who you ask, anything that could be cooked and eaten.[6]

Less alimentary and more seriously, Making Vinyl also featured a display from a Dutch group of companies operating under the banner of Green Vinyl Records. The company was there to demonstrate a disc made of a strange new secret material. Still, most of the conversations following the panel, in the foyers of the W Hollywood hotel and the bars of Hollywood Boulevard, focused on recycling oil-based plastics.

For example, an engineer from a Philadelphia pressing plant picked up on Dunster's comments about recovered ocean plastics. He talked in a detailed dialect of thermopolymers, focusing on the technical compatibility of mixed materials, like grocery bags recovered from the sea, with older and newer pressing machines. Although Dunster and others had found the task impractical, it was definitely possible to recycle such plastic, the engineer explained, handing me his card.

None of this was news to the next record presser I met, a tough-looking yet softspoken Canadian who was also exploring recycled discs. But his interest was not in ocean plastics. He was looking into old windows.

2
Vicious Cycles

Billy Bones plays in a punk rock motorbike concept band called the Vicious Cycles. Their sixth release, *Motorcycho*, according to one reviewer, is such a compelling tribute to the two-wheeled life that "you'll probably end up burning down your house and hitting the road forever." Billy also runs Clampdown Record Pressing near Vancouver, which strives to be one of the greenest record pressing plants in the world. To understand the company, it helps to understand the band. But first things first.[1]

Clampdown is short on vinyl. Everybody is. Orders are backed up. Small bands can't press small runs. Even some big bands can't press big ones. Part of the problem is a mysterious worldwide shortage of raw materials—and not just in the plastics industry.

But Clampdown's current delay is not that kind of mystery. Billy knows exactly where his shipment is. It's sitting across the border in the United States. He just doesn't know why. Maybe there is a customs hiccup. Or perhaps it is reduced trucker capacity. Whatever the cause, everyone at Clampdown is eagerly awaiting the shipment. Each time the buzzer rings, all eyes lock on the overhead delivery door.

Clampdown records waiting to ship from Vancouver, 2022.

To keep things going in the meantime, Billy is reviving some dead stock. He has recently returned from a trip up British Columbia's Sunshine Coast to collect a few boxes of unsold LPs from a local label. These can be reground and re-pressed into new records.

Regrinding has always been standard practice at Clampdown and throughout the industry—both upstream, at suppliers like Westlake Chemical, as well as downstream, at pressing plants themselves. But the practice takes on added significance in the face of climate crisis.

Billy is keenly aware of the problems of plastic pollution, and he knows what it means to make records from petroleum products. "People talk about vinyl and wax, and it sounds cool," he says, "but there is no getting around that vinyl records are PVC and every one that I make adds to the plastic in the world." This is another reason Clampdown regrinds, over and above the shortage. Clampdown even has a program encouraging bands to bring in their unsold records so they can be turned into new ones. It is one of the ways the company strives to be environmentally responsible—along with ditching shrink-wrap packaging, adopting post-consumer paperboard for album jackets, and using hydroelectric presses, which don't quite plug into a standard wall socket but do use less energy than the gas-powered boilers of traditional presses. In 2023 the company started planning to install a gas boiler, using biogas or renewable natural gas, as monthly bills for their electric water heaters were becoming unaffordable.[2]

Regrinding is essentially recycling, and, like all recycling, it is a laborious, conflicted process. One of the main difficulties surrounds the paper labels at the center of records. New discs do not sound good with paper in them, and because these labels are pressed into records during manufacture, not stuck onto them after, they cannot simply be peeled off. You have to cut them out.

Billy shows me the setup. In the back corner of the plant there is an old drill press fitted with a hole saw. An employee sits low to the ground on an orange office chair, flinging plastic shrapnel everywhere as he bores out the centers, one by one, disc by disc. Billy demonstrates. The label-less rings are then fed into a shredder, which breaks them into little black nuggets that are ready to be reborn as records.

The process is not just time consuming. It is also finicky. Because different batches of records use different plastic compounds, which means they have different melting properties, you can't easily mix dead stock from various record labels, nor can you simply amass piles of used records from miscellaneous flea markets, estate sales, and so on. Regrinding and recycling multiple-sourced records in this way would introduce too many unknowns and inconsistencies into the pressing process. It would pose problems for quality control. What's more, plenty of record lovers, including some Clampdown customers, believe the purest, freshest vinyl sounds best.

If the regrind process is time consuming and finicky, that means it is also expensive. Regrinding shares in the conflicts and challenges that mark recycling more generally. Because recycling operates as an industry, its aims are forked. It is directed toward environmental protection as its underlying principle, but it is also directed toward economic profit as its overarching priority. Ideally, protection and profit would coincide, and ideologically they do. In these scenarios, whether real cases or rational distortions, everybody wins. In situations where principle and priority are at odds, though, the imperative of market competition wins—and the environment loses. This hinders the recycling industry. It is the main reason the vinyl industry does not, as a general rule, regrind the billions of already existing records in order to prevent the production of new plastic.[3]

This is the reality for many pressing companies in the early 2020s. Vinyl is hard to come by, factories are at capacity, and regrind is only a partial answer. If ever there was a good time to make records out of something other than polyvinyl chloride, this is it.

Clampdown's founder has looked high and low for something other than PVC or at least a better way of using it. I ask Billy if anything ever came of the idea he had at Making Vinyl to make records out of recycled plastic window frames. Unfortunately not, he says, because windows start life as a different kind of plastic than records do. Window frames are made from unplasticized PVC, which lends them rigidity and durability, meaning they are long lasting and low maintenance from a construction standpoint. Unplasticized PVC is not suitable for records. Low maintenance in construction does not translate to high fidelity in music.

Billy was corresponding with a Canadian business that was making plastic products from reclaimed lumber. This seemed promising, though for unknown reasons the company stopped returning his calls. He wondered too about the potential of corn-based bioplastic, observing the prominence of cutlery made from compostable vegetal materials. Billy even has a caution-to-the-wind vision of ignoring the difficulties of mixed regrind—of doing a thrift store road trip, gathering old records, and re-pressing them anyway. Maybe the quality will be okay. Maybe the next Vicious Cycles record will be made in this way.

For now, though, Billy Bones is in a bind. He can't get vinyl. But he can't get away from it either.

Billy lives another, related bind. He runs a company, sometimes around the clock. He has investors. He has employees. He is a businessman. Billy makes no bones about this. Still, he is not an enthusiastic capitalist. He is an anarchist with syndicalist sympathies and sometimes socialist leanings too. But the world is the world, he tells me, and he has people who depend on him. He has to keep the lights on. And, as he works to do that, Billy searches, as we all do, for meaning and purpose in an economic arrangement that accommodates—and, in a dark irony, both produces and profits from—these tensions between political ideals and practical realities.

After the factory tour, over cheeseburgers and soda, under a gray sky in a parking lot, we talk about what we're listening to these days. I'm lost in Loretta Lynn and Roger Miller. Billy is also listening to sixties stuff, like girl groups and reggae, and he says the most recent album by Personal and the Pizzas, another garage-ish concept band like the Vicious Cycles, is one of the best releases in the past fifteen years. We talk too about how our worldviews were formed by the music we heard at impressionable ages. For both of us, this means punk.

Probably the most common observation about punk is that it embodies tensions and registers contradictions. Whether writing five decades ago or today, the genre's most insightful commentators have always observed how punk fuses rebelliousness with commodification, capitalism with resistance. The Sex Pistols were dangerous; the Sex Pistols were a boy band. Arguments like this go back and forth and round and round, both sides—both truths—defining, aggravating, intensifying one another. But this cycle is not vicious. It is a tractive friction, the defining feature not just of punk but of most music and culture since industrialization.[4]

The same basic friction defines technical responses to climate issues. Our burning desire to do whatever we can to help solve the crisis can be satisfyingly quenched within the confines of the economy that has itself begotten the crisis. For better and worse, this situation is the source of a beauty Clampdown embodies—the same beauty I find in music's recomposition more generally. It is also the source of what an early punk observer once called a strangely neglected topic of social inquiry. Fun.[5]

Unable to resolve the dilemmas, with the politics of resistance and resignation spinning in my head, I retreated to my hotel and put on *Motorcycho*.[6]

I wanna ride
I wanna ride
I wanna ride around

Let's go get hotdogs in the city
Let's go play pinball all night.

3
Green Vinyl

Back in Hollywood, the vinyl traditionalists are needling that strange new record. Its ordinary look disguises the disc's divergence from a typical LP, in terms of both process and substance. This record is molded by injection rather than compression, which saves energy and reduces pollution. It is made from something other than polyvinyl chloride, something less toxic and more recyclable. With all that is known about PVC—fossil fuels, material shortages, price hikes, looming bans—this novel record seems to be what everyone is looking for. It should be a no-brainer.

But it sounds awful, they say. It feels cheap. The alien disc, according to the conference chatter at Making Vinyl, is nothing but a gimmick.

Although I do not share their jealous devotion to the LP, the vinyl buffs have a point. This thing neither feels nor sounds like a typical record. The flex is unusual. The edges are boxy. And the background noise is unique. It doesn't sizzle like shellac, pop like vinyl, or fizz like tape. But it does whine sometimes. Still, the differences are minor. I could get used to them. Probably others could too. Because if these are the features of a record that is less damaging than a traditional LP, such tactile and acoustic idiosyncrasies are not blemishes. They are beauty marks. From this perspective, the vinyl traditionalists have it all wrong. It is the familiar warmth of the regular record, an index of the fossil fuel and PVC industries, that actually sounds awful.[1]

Harm Theunisse doesn't listen to the vinyl industry's old guard anymore. He doesn't have to. Theunisse's company, Green Vinyl Records, has come a long way since Hollywood. He has urged me to visit his operation in the Netherlands to see for myself.

Green Vinyl's research, development, and production all happen about a half-hour drive east of Eindhoven. In terms of the history of recording, this is hallowed ground. We're in Philips territory, longtime headquarters of the company that developed the cassette tape and the compact disc.

Theunisse knows a thing or two about CDs. He got his start with them. As the format was beginning to peak in the 1990s, Theunisse worked on the floor of an EMI disc manufacturing plant in Uden, north of Eindhoven. He soon struck out on his own, and today, with his brother, he runs Symcon.

Symcon is a full-range digital optical media company (making not just CDs but also DVDs, Blu-ray discs, and more) with manufacturing facilities in eighty countries as well as offices in the Netherlands, the United States, Brazil, Columbia, Taiwan, and Hong Kong. It is a point of pride for Theunisse that Symcon eventually bought the company that supplied plastic to EMI when he worked there all those years ago.

CDs and Symcon are the body and soul of Green Vinyl. The project itself, though, began, in the way these stories do, as scribbled notes on a pack of cigarettes—notes of disbelief that records were still made by compression molding, like waffles. Injection molding, which is more like sausage stuffing than waffle making, has certain advantages over the compression method. It is also the procedure used to make digital optical media such as CDs. Naturally, this seemed far more sensible than compression to both the CD-savvy Theunisse and his eventual colleague Pieter van Ettro, Green Vinyl's resident audiophile and a retired injection engineer.

In the early-to-mid 2010s, with vinyl record sales clearly on the rise, the pair saw an opportunity to put their injection molding and optical media expertise to use in a new market. Nearly a decade later, with some one-time innovation funding awarded by the European Union plus the support provided by several partners, including Symcon, Green Vinyl is still in a phase of research and development, not yet fully commercially viable. But it is close.

When I visit, Theunisse is perfecting the job of printing labels directly onto the discs. It is the last step in the manufacturing process and the last technical hurdle before he can start making discs commercially. The idea is not so different from what happens inside a color inkjet printer. In terms of recycling, printing has a major upside. Unlike standard LPs, there is no pesky paper label—and so no costly removal process. This is one reason Green Vinyl promotes its greenness. If you wanted, these records could go directly into your household recycling bin.

Another important factor is what the discs themselves are made of. This is the source of some of the mystique surrounding Green Vinyl. Is the product made from a non-petroleum substance? Understandably, in the past Theunisse has been reserved about this information. It is part of their proprietary process. But he is open about it now, and anyway the 1,200-kilogram flexible intermediate bulk containers inside the Green Vinyl factory are hard to miss.

These FIBCs are like giant, industrial-scale versions of those blue IKEA bags. They confirm that Green Vinyl is making its records from polyester terephthalate, or PET. I ask if they use a special mixture of polyester that includes various additives, also known as a plastic compound, and am told they use pure PET resin—albeit with a tweak early in the polymerization process, making it suitable for molding records by injection.

Polyester terephthalate is common. Odds are there is some of it in the socks you are wearing, the pop bottle in your fridge, or the insulation in your walls. Polyester is also far from innocent. It is a petroleum product, and it is a major part of the microplastic pollution problem in our oceans and our bodies. But it does have the advantage of not being PVC, which has long been recognized as "the most environmentally damaging" type of plastic. Here too, compared to traditional records, Green Vinyl has the upper hand.[2]

The final main environmental advantage of these records has to do with injection molding itself. Theunisse's analysis shows that his process is more efficient than compression vinyl. This is largely because he does not use a gas boiler to power his machines, which is the industry standard. Making a Green Vinyl disc uses the amount of electricity it takes to shine an everyday sixty-watt lightbulb for around three hours. A standard LP would be closer to fifty hours of lightbulb use. With an estimated production run of 200 million discs per year (which shows the scale of ambition at Green Vinyl), Theunisse projects energy savings of more than 90 percent.

Production involves an amalgamation of four machines working together in automated synchrony, machines that Theunisse and his engineers have custom-built to some degree. There is the injection device itself, smaller than a cargo container but bigger than an SUV, adapted from technology originally designed for the LaserDisc—which, not coincidentally, is an optical media format with the same

Green Vinyl machine in action near Eindhoven, 2021.

width as the LP. There is the rotating cooling rack, precision-machined and spinning like clockwork. There is the label printer, running on seven different computers coded together. And linking them all, with precision and grace, is a robotic arm that would be more at home in an ultramodern automobile plant than a record factory.

I ask Theunisse how he learned to do this kind of work, given that he has no formal training as an engineer or computer scientist. "Farmer's intuition," he says, directly translating a Dutch expression meaning something between folk wisdom, common sense, and street smarts. Only employees at Green Vinyl can operate their creation. But the idea is to make it streamlined, automated, and foolproof, so that it requires less specialized knowledge to run—similar to the difference between piloting a fighter jet and piloting a passenger plane.

This is important, as Theunisse's business plan does not stop at manufacturing records. He also wants to sell the machines and license the process, a bit like a franchise.

It is all truly impressive. But how do Green Vinyl records sound?

Theunisse and Van Ettro pull a randomly selected disc hot off the mold, insisting that I be the only one to handle it as we walk to the listening room. It feels a bit over the top, like a card trick, but I am assured it is necessary. Skeptics have accused them of cherry-picking the best examples of their product, even swap-outs by sleight of hand. The showmanship ensures we will listen to a true sample.

The disc sounds perfect, certainly as good as anything I've listened to on record. Not that I have golden ears—but I do not need an especially refined auditory palate in this scenario. Everything is right there in front of me, on digital display. Green Vinyl discs have a noise floor and frequency range that outperform those of traditional LPs. Under a special light, you can see residual stress marks in a compression record, like ripples on a pond. No such birefringence can be detected in Green Vinyl's discs, which is because injection is less violent than compression. These records also have greater durability. I am told they can be played 400 times before exhibiting any noticeable wear and tear. This compares to about seventy plays on a standard LP. Plus, thanks to their paperless ink labels, Green Vinyl is not just fully recyclable. It is also dishwasher safe.

That evening, Harm Theunisse and I let our imaginations roam regarding the future of Green Vinyl. The discs sound great. The timing is right. Once the label printer is at peak performance, there is no technical reason these records shouldn't take over the world.

History is on Green Vinyl's side—and not. On the one hand, even if Green Vinyl was imperfect from an audio standpoint, the entire history of commercial sound reproduction is a story of various practical considerations winning out over so-called high fidelity. Around 1900, horns sounded worse than ear tubes and discs sounded worse than cylinders. But horns and discs were louder, meaning better for dancing, and so were preferred by listeners. Early MP3s too could be poor in terms of audio resolution. But the format offered new kinds of storage and portability, which were considered fair tradeoffs for sloshy sound. Knowing the history of sound reproduction makes it easy to imagine listeners today choosing a green disc before a less green one, almost regardless of what it sounds like. But this is not something anyone needs to imagine. Theunisse and Van Ettro have worked hard to prove themselves to the audiophile gatekeepers. Green Vinyl is the acoustic equal of its peers.

On the other hand, though, the blessing of history is more mixed. This brings us back to cigarette packs and disbelief. Theunisse's and Van Ettro's surprise that twenty-first-century records are still made by compression is understandable, given their embeddedness in the world of injection molding. And it is true that LPs have never been mass-produced by injection. But that is not for lack of trying.

Injection molding has always been touted as the future of records. It has been there from the start of music's plastic era. Already in the early 1950s, *Billboard* called injection molding "the most spectacular advancement in modern record production," while Columbia was making 90 percent of its 45 rpm singles and EPs by injection as well as investing millions in the hopes of improving their injection-molded LPs. The challenge with LPs was not injection itself. It was the pairing of process and material.[3]

Neither vinyl nor shellac, the record industry's earlier staple commodity, was suited to injection molding. Those materials were confined to compression. Polystyrene, by contrast, worked well for injection. It was widely used for children's records, or kidisks, as well as seven-inch discs like 45s. With styrene-injected LPs, however, reviews varied. Columbia claimed its injection LP was "a better sounding disk than the compression LP," which makes sense given the company's investment. Those firms with less buy-in, though, including "other majors, and many indie pressing plants," were "not yet convinced"—stymied by styrene. Almost a decade after the 1950s push for styrene and injection, "most engineers preferred vinyl for top quality sound reproduction." The same is true today.[4]

Still, the record industry never stopped pursuing injection. Done right, disc makers always believed, the advantages would be significant. They foresaw not just potentially better sound but also "lower material costs, lower labor costs, and higher production rates." With the exception of the ecological angle, which only recently became routine marketing, most of the injection advantages described by Green Vinyl have been known—and chased—for seventy years.

Some early engineers were confident about the future. "Even if injection LPs are not as perfect now quality-wise as compression LPs," one said, "it is only a matter of time until the injection LP is the equal of the other." Maybe that time has finally arrived.[5]

Although injection has long been on the horizon of the recording industry's future, there is another area where it is decidedly passé. Media studies.

Contemporary media researchers are not especially curious about whether media content can be transmitted hypodermically, either directly (from sender to receiver) or uniformly (from reality to representation). They are more likely to ask "how media manage and enact relations shaped by one or more conditions of finitude." This way of thinking has opened adventurous possibilities for writing media history sideways, through nonstandard general categories like clouds, sky, water, waves, heat, saturation—and, yes, compression. It is not clear whether a general media history of injection would lead to anything so interesting. But the possibility should not be dismissed.[6]

For now, it is enough to understand Green Vinyl through an existing truism in media studies. The content of a medium tends to be another medium. Speech is the content of writing, theater the content of radio, radio the content of television, LPs the content of CDs—and, in an ordinary turn of remediation, the CD is the content of Green Vinyl's LP. This does not mean that people can hear the circuitry of the CD in the grooves of this LP. It is more that the historical reception of the compact disc—figured as cold, scientific, and disenchanted against the warm, romantic, and enchanted LP—lurks in the initial scoffing response to Green Vinyl among the traditionalists.

CDs are in this LP's genes. Search its soul, though, and you find Symcon—the ethos of big business. This differentiates Green Vinyl from a specific aspect of the resurgence in the popularity of records. While the so-called vinyl revival's demand is certainly being supplied by major facilities, such as United Record Pressing in the United States and GZ Media in the Czech Republic, the phenomenon has been accompanied by a proliferation of smaller-scale, craft-oriented, independent pressing plants from Austin to Auckland and beyond—Clampdown included. The keyword used by many of these businesses to describe themselves is *boutique*.

Record pressing is not alone here. Against the backdrop of economic restructuring going back to the 1970s, as certain countries became oriented around service and knowledge rather than manufacturing, and especially since the global financial meltdown that began in 2007, some parts of the population have reassessed and reassumed what were once considered undesirable jobs or unthinkable vocations. These undertakings tend to emphasize hands-on, self-sufficient, small-scale forms of work. Sometimes this takes the form of beekeeping or bartending, in which case it is described in terms of craft entrepreneurship and artisan economies. Sometimes it takes the form of meth cooking and drug dealing, in which case it is discussed in terms of illicit alchemy and informal economies. Always it promises a more fulfilling life for the people hit hardest by deindustrialization.[7]

Boutique records take their place alongside artisanal apiculture and mixology as well as today's boutique butchers, brewers, bakers, and more. As the economist Adam Smith once wrote, we can expect our dinner from the latter group not because they are benevolent but because they are self-interested. If the market can pull together competing self-interests to prepare a meal that no individual intends, then what is stopping it from providing ecofriendly goods simply by coordinating artisanal, sympathetic producers with their sympathetic consumers? Maybe the invisible hand really can have a green thumb.[8]

It might be tempting to leap toward the seemingly obvious conclusion about Green Vinyl—that, despite the possible ecological benefits offered by its product, the company's big-business ambitions are antithetical to environmentalism. It is true, for example, that, as companies and industries grow larger, technical improvements are often glued to rebound effects. This means that Theunisse's estimated 90 percent efficiency gain over the standard LP method could be undone by an overall increase in production. The math is easy.

If making one Green Vinyl record uses the same amount of electricity it takes to shine a lightbulb for three hours, and a standard LP uses about fifty hours, then the efficiency gain starts to backfire if you make more than fifteen times the number of Green Vinyl records as the standard method. At scale, this is a lot. Were the US industry to sell 40 million LPs made by the standard method, which it did in 2021, the industry would be able to produce over 600 million Green Vinyl discs before exceeding the technology's efficiency gains. To meet Theunisse's aspiration of 200 million records per year, Green Vinyl would be capable of producing more than 3 billion records before starting to see a rebound effect over polyvinyl compression LPs.

Although such sales figures are far-fetched, they cannot be ruled out. But the boutique world—the artisan economy, the new spirit of capital—has its own built-in contradictions and limitations, especially when it comes to the climate crisis. One issue is that such formations pose no real challenge to our economic arrangement, the source of the problem they profess to address. In fact, according to sociologists, "the artisan economy ideologically legitimizes capitalism while also contributing directly to accumulation and reproduction." Artisanal production and boutique products simply express a preference for a particular kind of capital: mom-and-pop shops as opposed to big-box stores. Which leads to another problem with such business models. They will not scale.[9]

This is by definition and design. These businesses represent a kind of small-is-beautiful retreat from scalability and mass production. The products of such businesses tend to be crafted with great care, in small runs, and in ways that foreground traditional techniques while showcasing their supply chains. They are therefore expensive—and exclusive. Such products have a tendency to become status symbols, wrapped up in the forms of sympathetic and conspicuous consumption that are available to those with particular sensibilities and incomes. From an ecological perspective, while Green Vinyl's business model may risk backfiring somewhere down the line, the boutique business ethos risks, from the get-go, amounting to little more than an environmentalism of the rich.[10]

In the struggle for the planet, boutique localism will not do. Whether dealing with music—or more basic needs like food, shelter, and electricity—the fact of the matter is that most people in most places need more of these things, not less. Most people need production scaled up, not down. This is true not only along the North–South axis that defines globally unequal exchange. It is also true along the upper and lower class axes that define social orders in both the North and South themselves. By contrast, many well-meaning, climate-concerned citizens in advantaged social positions are energized by a politics of less. They are interested in cultivating philosophies of post-growth living. Some of this may be required. But such ideas are connected to the small-is-beautiful, artisanal, boutique values that articulate a vision of social reality (and social change) that can only be understood as emanating from the contradictory class location that produces those values in the first place—and that purveys the means to pursue them.

Large-scale production, like what is envisioned by Green Vinyl, is required. It is not automatically antithetical to environmentalism. More precisely, if large-scale industrial projects are antithetical to a particular and widespread environmentalist sensibility, they are not necessarily antithetical to serious efforts at saving the planet or building a more just world. Arranging economies in the service of capital is what gets in the way of those goals.

Green Vinyl, as with the record industry and much else, is beholden not just to capital. It is beholden to a specific form of capital that has been built around fossil fuels. There are many ways that our social and cultural worlds have been made possible not just by the energy density and portability offered by coal, gas, and oil but also by the way those fuels have been put to work by fossil capital.

Fossil capital has created a world that is physical and material, of course, but also one of shared values, beliefs, and customs. In other words, fossil capital and its petroculture have shaped things like how we get around as well as what we value about mobility, how we listen as well as what we value about music. Fossil capital is the energetic chicken that laid the social egg from which it hatched—and this strange creature has come home to roost. The question is what music (not to mention the rest of the world) would be like if that cyclical dependence was cut off and replaced by something else.[11]

Answers are emerging. A company out of Britain, for example, although sticking with compression molding, is abandoning oil. In doing so, they are giving new meaning to the phrase *pressing plants*.

4

Evolution Music

"It's happening," Steve Charter says. "It's happening."

We're at a bar just east of Frankfurt, in the August heat on the eve of Making Vinyl 2022, and news had broken just hours earlier. In two days, Michael Stipe and Beatie Wolf will release a split twelve-inch single. But this is no ordinary record. "The disc will not be pressed on traditional polyvinyl chloride," wrote *Pitchfork*, one of the first to cover the story. "Instead, the record will be the world's first commercially available twelve-inch made from sustainable bioplastic." It is the greatest potential change to the substance of the disc format since vinyl replaced shellac around 1950.[1]

For Charter, the release marks the culmination of four years' dedicated work with Evolution Music, a "collective of music industry and sustainability experts" that has researched and developed the bioplastic record. Equally, the release can be seen as a high point in a lifetime devoted to environmentalism.

Charter has been involved with sustainability since the 1990s, with projects that have embraced principles of circular economics, a kind of cradle-to-cradle model of production and consumption, as well as permaculture, a naturalistic approach to agriculture. His activities in these areas have ranged from urban planning and housing projects to construction training and consultancy, all of which influence his current role as Evolution's cofounder and director of sustainability. Charter even has a soft spot for permaculture music, featuring bands like the Formidable Vegetable Sound System.

Evolution Music thinks big. They have ambitions to introduce principles of sustainability throughout the music industry, including a carbon-neutral record pressing facility. The bioplastic record, which they describe as their "initial flagship product," is only a first step in those larger plans. But it is an important one.

The discs are made of a polymer called polylactic acid (PLA), which itself is made from plants instead of the petroleum used in PVC. It is also renewable, recyclable, and biodegradable. This is promising from an environmental perspective. From a business perspective the promise has been less obvious. The difficulty for Evolution is that PLA exists. It is already widely used in disposable cups, cutlery, plates, food packaging, 3D printing, and more. You can't patent polylactic acid. If Evolution Music simply brokered PLA, there would be nothing to prevent a record company or pressing plant from going directly to a supplier and purchasing its bioplastic wholesale, bypassing the company entirely.

Charter and his partners had a plan. Their initial strategy, according to music industry veteran and Evolution Music CEO Marc Carey, was to make a claim on PLA by laying a defensive legal minefield around their work. This included trademarking terms such as "the original bioplastic LP" and "music made better," publishing a variety of "first" claims in the media, and inking certain supply chain contracts. Carey describes this as a kind of Kentucky Fried Chicken model. Sure, there is fried chicken on every street corner. You can't patent that. But you can build a brand, a trademark, and a series of exclusive supplier agreements—which allows you to market and sell a specific product.

Evolution Music had their KFC model in place. However, as they discovered, it is not easy to develop a new material that meets the performance and manufacturing standards of a product whose feedstock has undergone three quarters of a century of fine tuning. While this posed numerous challenges in terms of R&D, it also presented an opportunity. It became necessary to modify the standard form of PLA bioplastic resin.

The work was slow—and then fast. Having incorporated in 2018, by late 2021 Evolution Music was making their first test pressings. The discs were of sufficient quality and intrigue to attract the attention of a respected vinyl plant, which scheduled a follow-up test in January 2022.

This second test was a disaster. There were complications due to lockdowns, which meant that no one from Evolution Music or Colloids, the Liverpool-based supplier of their experimental PLA mixture, was able to observe the process. Reports came back that "no good pressings were achieved" and that "the product was extremely runny and difficult to work with" on the record presses. Through subsequent analysis, Evolution determined that the bioplastic mixture used in this test had been contaminated.[2]

The next attempt took place in May at another well-known vinyl manufacturing facility. This time things went well, and the discs formed correctly in the presses. While it was true that the surface noise was slightly more noticeable than on a traditional LP, further studies traced that problem "to the material composition rather than the manufacturing (pressing) process." There was consensus on the next step. "All involved believe that the surface noise issue will be resolved by fine tuning the mix with replacement organic fillers."

Evovinyl bioplastic records, 2025. Photo by Evolution Music.

As they continued testing several varieties of "amended compound," which included a chalky substance in the master batch (a mixture of additives that lends certain properties to pure plastic resin), Evolution stepped up its testing. Working with companies like Press On Vinyl in Middlesbrough, Vinyl Presents near London, and Deep Grooves in the Netherlands, they have been able to produce prototypes that, according to Carey, are about 95 percent similar to a standard LP. Listening to my test record, this sounds about right. If you pay close attention, there is still some surface noise. But it's not very noticeable—and, anyway, it is a fillip for moving beyond fossil capital. Nonetheless, Evolution is striving for that last few percent. In the meantime, this is the bioplastic iteration that Stipe and Wolf released in a limited run of 500 copies, which sold out in six hours.

What all this means is that Evolution no longer has to rely on the KFC model. They have invented a PLA cocktail of their own—Evovinyl™ (patent pending)—that is specifically designed for records and engineered to work on the existing equipment of vinyl pressing facilities. They plan to sell this patented compound wholesale to pressing plants. The bioplastic disc, a new era in music's ecological record, seems to have arrived.

It really is happening.

As Charter updated me on various other developments at Evolution Music, the 2022 Making Vinyl welcome reception started picking up. There were representatives from established companies like Optimal Media, a large German manufacturer of everything from records to books, as well as newer ones, including ResurRec, a pressing plant in Bangkok. There were companies reviving old formats like singing postcards (records you can put in the mail) and startups making discs from recycled medical equipment packaging (ocean plastics proved overly brittle). Harm Theunisse was there too. He was in good spirits, as usual—and very busy, as usual. Theunisse had started taking commercial orders from major labels even as he continued to reinvent Green Vinyl's label printing process. The whole reception had a familiar air of hobnobbing and reunion, like good parties do.

Yet the overall atmosphere had changed from that of the Making Vinyl event I presented at in 2019. There was still a contagious optimism about vinyl, as record sales were soaring. But several discussions and presentations also focused on bottlenecks in production, including shortages in plastics, plating, and labor. Panels had titles like "Sorry, All With Vinyl Is Not Rosy (Three Contrarian Viewpoints)." Industry insiders and analysts forecasted that the seemingly unstoppable two-decade escalation in vinyl sales, which led to the new wave of boutique pressing plants, would soon stall—and then fall. There was a shift in sustainability thinking too. By 2022, optimism had outpaced the modest idea that vinyl was "more sustainable than you think." Now the thinking was represented by another panel title: "Sustainability Will Sustain Vinyl's Future."

Not everyone was buying it, though. On 2022's sustainability panel, for example, Lupu Pitkänen of Helsinki Record Pressing (Puristamo) questioned the industry's preoccupation with discs. His point was that energy use in the production process was a bigger problem, which is why Puristamo avails itself of renewables to power its presses. It is also why Pitkänen has devised a system whereby any heat energy that goes unused in production can be recovered and sold back to the energy grid, powering a number of homes in Finland. The point is sensible. In fact, it is the reason Billy Bones and Harm Theunisse focus as much on the production process as the disc materials. The same thinking underlies Evolution's desire to change how pressing plants work.

During "Meeting Consumer Demand for Vinyl When the Industry Is at Capacity," things got a bit tense when someone suggested that newer materials such as bioplastic could alleviate the holdups and doubts surrounding PVC. A director from a major vinyl manufacturer was particularly defensive. He believed that PVC's reputation as "the most environmentally damaging" plastic was unfair, and he insisted that there were benefits to PVC that deserved emphasis. For example, he noted the recyclability of vinyl and his company's use of regrind. He also noted that there have been advances in using nonpetroleum feedstock chemicals in the polymerization of PVC. In addition to making these points publicly, the director had taken the time to outline some of this to individual delegates, offering a series of pointed questions about both the viability and the real ecological benefits of bioplastic.

It was in anticipation of reactions like these that Evolution's Marc Carey feared he was "walking into a lion's den" at Making Vinyl, with regard to both the sustainability panel itself, on which he spoke along with several others connected to the company, as well as Evolution's presence at the conference more generally. The overall acceptance and enthusiasm came as pleasant affirmations, Carey said, if not complete surprises. But it was actually a form of indifference in another setting that cemented the company's confidence in its product.

Carey told a story about meeting a record industry higher-up to demonstrate the bioplastic record. After all the work, all the testing, all the improvement, all the investment, it was a big moment. There was a sense that Evolution's future hung in the balance. The record executive's determination was swift and definite. "Looks like vinyl. Sounds like vinyl. When can we get it?" For a format aiming to be vinyl but better, that was high praise.

With the challenges of invention and quality mostly out of the way, attention can turn to another question, one that comes up often when discussing Evolution Music, in shades of both enthusiasm and skepticism. It is the question that the defensive vinyl director and the matter-of-fact record executive were really getting at. Will it scale?

5
Will It Scale?

Scalability in manufacturing usually comes down to whether producing more goods will result in lower costs, an advantage known as an economy of scale. There is also a related idea in business: scale thinking. In this sense, scalability means "the ability to expand—and expand and expand—without rethinking basic elements." Though an old idea, scale thinking has been the focus of much business attention in our own time, especially in Silicon Valley. Scale thinking has also been used to characterize social problems more generally—as well as any potential political interventions with regard to those problems.[1]

For Evolution Music, all these considerations are relevant. Can their small-batch test pressings be scaled up from a few hundred discs to millions? Will doing so be cost-effective? And is it possible to do so in a way that follows a template permitting delivery of the same basic product at a volume capable of matching market forces? Assuming the answers to these questions are yes, the crucial question is what social or ecological effects such scaling may have.

The business world may advocate scalability on all fronts, and political life may increasingly be defined by scale thinking. But researchers have looked upon scalability with suspicion.

Anthropologists, for instance, suggest that the practice of scalability precedes the term, noting the historical coincidence of scalability and sugarcane in webs of slavery and colonialism in the seventeenth century and after.

Here scalability follows an expansionist blueprint that assumes control of land, establishes standardized infrastructures that destroy local fauna, flora, and fungi, and then supplants them with cash crops tended by enslaved people. In these ways, the pursuit of scalable world-making projects tends to leave behind "mounting piles of ruins." Scalability is an enemy of the humane, a naturalizer of the unnatural. It replaces "relations of care" and "banishes meaningful diversity," conditions that, for some anthropologists, are most meaningfully located in smaller-scale worlds, life-forms, and relationships.[2]

This portrayal is not wrong. History shows that plantation sugar's politics of scale are bound up with the devastation of cultures and environments—at least as much as any other colonial crop. Which is all the more relevant given the particular plant matter from which polylactic acid, and Evolution Music's bioplastic LP, are derived. Sugar.[3]

It is true that, with sugar as a precursor, Evolution Music's Evovinyl may not enact the fossil fuel economy, at least not directly, in the disc itself. (Fossil fuels are probably still used throughout the polymerization and pressing processes. They will be unavoidable in shipping feedstocks and finished products.) This ecological record may boast real benefits at the level of carbon emissions and plastic pollution. But the bioplastics boom and the large-scale expectations that listeners today bring to the music industry and cultures of consumption present a conundrum. Evolution aims to accommodate these expectations, but if sucrose records reanimate the politics of scale that oversaw sugarcane colonialism (and continue today on a spectrum of sugarcane imperialism), then how much progress will have been made? It is a question Evolution Music is asking openly and earnestly.

For scalability skeptics, the proposed remedy for rethinking and remaking the world is non-scalability—projects that emphasize "the transformative diversity of economic niches." Certain small-scale or non-scalable projects will be necessary and beneficial in meeting the current challenges of life on Earth. Equally, though, the trouble with the climate crisis is precisely that the threat is not being met at the scale (or with the speed) that is required.[4]

It is the skeptics' own logic that lends hope and political vision in these terms. If their guiding caveat is right—that "nonscalability is by no means better than scalability just by being nonscalable," that "both good and bad things can be nonscalable"—then the opposite also holds. Scalability can exist for good and ill. As with Green Vinyl, certain types of scaled-up projects will be fundamental if the climate crisis (and its underlying realities) are to be addressed in ways that do not simply repeat the inadequacies and biases of non-scalable, luxury environmentalism.[5]

In the music world, Clampdown, Green Vinyl, and Evolution Music all define their own problems and solutions on their own scales. The effectiveness of their interventions should be gauged not primarily on the basis of their products, which would be to repeat the determinist error of technical solutionism, but rather in terms of the social realities and the utopian horizons encompassing the scalable and non-scalable relations that are objectified in their products and business models.

In the end, that's what scalability is: a form of objectification. It is a way of formalizing, externalizing, and regularizing the relationships that are deemed essential to producing those goods and services required to maintain social orders across times and places. No more, no less. In other words, objectification is a "metaphysical process, by which socially negotiated relationships between people are made to appear as relations between objects, and hence as objective." Objectification is therefore about things and people—but it is also about the ways that people make things think for them and manage their affairs. Like scalability, then, objectification is neither bad nor good. It is what we make of it. And we are what it makes of us.[6]

This is to take a political position on objectification that may be counterintuitive in certain schools of thought. These schools are sometimes referred to as critical, on the one hand, and pragmatic, on the other. In the critical approach, objectification is constituted by estranged human relationships. Objectified relations tend to be contorted relations, which represent themselves in the form of objects that obscure the fact that there are human connections and histories underlying all this in the first place. The tendency in this form of critical analysis is therefore toward demystification—stripping away the facade of objectification, bringing what has been estranged back into the realm of the familiar, in order to see what's really going on.

Demystification is enjoying a moment of prominence both in contemporary musical life and well beyond it. Numerous publications and advertisements now follow the supply chains of various musical commodities. Instruments are traced back to trees, electronics to mines and factories. Similar impulses define many sales pitches these days. The guiding conviction seems to be that if only we could unveil, once and for all, the truth of the reality behind the mask of mystification, such knowledge would be powerful enough to instigate social change. The limitations of this vision are multiple.

It is known that demystification, in and of itself, is ineffective as a means of altering how the world works. Such thinking is tied to a centuries-old history of framing consumption politically and doing politics consumptively—histories of sympathetic production and consumption that, by and large, keep failing as mechanisms of improvement or change. In fact, this kind of demystification is itself a popular form of entertainment that is entirely amenable to capital. Think about shows like *How It's Made* and *Dirty Jobs*, along with countless other programs and YouTube channels.

Demystification can also take the form of supply chain showcasing, which is a means of corporate promotion going back at least a century in major advertising campaigns and has attained new prominence with today's boutique products and artisan economies. This should give pause. Why are these appetites so central to both commodities and entertainment as well as contemporary social inquiry? How might those projects coincide?

In terms of addressing climate or ecological issues, the real trick is not that the mask conceals things. It is that there is nothing behind the mask after all. The critical impulse here takes on the appearance of demystification even as it deposits an additional layer of mystery onto commodity culture. Despite the best intentions, such tacks can end up creating "a distortion of reality which reifies and reproduces the fundamental process of capitalism by making the commodity form the solution to its own mystifications." They can embody the "predicament of a social formation that offers its agents the means to reproduce its own structure while simultaneously feeling as though they are toppling it." None of this is to say that demystification is fruitless or backwards. It is simply to emphasize that demystifying objects is a starting point for social inquiry,, not its final destination.[7]

The more pragmatic school of thought treats objectification as a social fact. Rather than trying to reveal what is going on behind the scenes, research in this arena is more interested in what objects make possible on the stage of social life. Some of these researchers have been skeptical about the critical school of thought for referring to the way people interact with their surroundings in terms of misrecognition or false consciousness, implying that everyday experiences and appearances can be deceiving. Instead, their focus is on how "users matter" in the social construction and implementation of technical objects—as well as how technical objects may have "agency" or may "afford" certain possibilities, more and less readily, in localized encounters between people and things.

This perspective underpins certain prominent strains of science and technology studies whose researchers consciously oppose themselves to so-called critical research. As with the enthusiasm for demystification, such strains run through the contemporary music world. They are in the water, so to speak. But that does not mean we should simply let them wash over our thoughts.

It is no accident that the pragmatic perspective is connected, by a fairly direct historical line, to the implication of anthropological and sociological ethnography in the world of information technology, and human-computer interaction in particular, especially beginning in the 1970s at the Xerox Palo Alto Research Center. This was a moment when marketing firms began turning from individual psychology to cultural context, and when technology designers started turning from the "particular features of a computer" to "the detailed situational use of the technology," in their attempts to understand consumer behavior, design better products, and ultimately constitute bigger markets. While the benefits of such work can be real, as feminist and disability researchers have shown, its attendant political vision, or utopian horizon, by focusing on objects as such, has also been criticized for effectively paralleling or consigning itself to market research.[8]

Thirty years ago, political theorists questioned this version of science and technology research for disregarding "the possibility that there may be dynamics evident in technological change beyond those revealed by studying the immediate needs, interests, problems, and solutions of specific groups and social actors." Research in the pragmatic vein frequently misses, or dismisses, the fact that what is most materially consequential in social life is not what is most concrete (objects) but rather what is most abstract (systems of objectification). It is often those things that may not be immediately accessible to experience or empirical description that exercise the strongest agency in our lives.[9]

As various individuals and companies work to change how records are made, for the benefit of the climate, it is worth trying to think differently about objects and objectification. Because "if we seize only the means of producing the former, we will, in the end, only reproduce the logic we hope to escape." This is the tendency for scratches to turn back into the itches they set out to relieve.[10]

Here we arrive at a structural mismatch between those who produce objects and those who account for objectification. Smaller-scale producers, like Clampdown, may have the inclination to question how objectification works according to the logic of capital. Yet they only have the power to influence objects themselves. Larger-scale producers, on the other hand, like Green Vinyl or Evolution Music at least as they aspire to be, potentially have the power to influence objectification. Yet the logic of capital means they only have the inclination to influence objects. These are the tensions between economic priorities and ecological principles that define climate crisis under green capital.

There is another way. It involves neither accepting objectification as a social fact and studying its effects nor rejecting objectification as a social illusion and dispelling its mysteries. Instead, it is possible to try improving objectification, by learning to build better fetishes.

6
Building Better Fetishes

Records hold a special place in the imaginations of performers and listeners. Musicians and fans are still investing in records, financially and emotionally, even though there is no strictly technological reason that they should be. Countless academic and journalistic essays comment on this fact, searching tirelessly for what it all means, how it all works, and where it all comes from. What interests me is the prominence of a particular way of describing this format, the attachments people have to it, and its role in the music world. Records are talked about not just as desirable musical things or important sonic objects. Loving records is said to be a kind of fetishism.

In a book about where stuff comes from, Harvey Molotch writes: "How we desire, produce, and discard the durables of existence helps form who we are, how we connect to one another, and what we do to the earth." Looking at everything from computers to toasters, Molotch wagers, "We can improve goods—both socially and ecologically—and comprehend more about the society that produces them if we understand how and why they come to be as they are."[1]

He's right, and one goal here is to describe how ecological records are coming to be as they are. But it is also possible to move in a complementary direction. My bet is that we can improve the world a little—both socially and ecologically—by building better fetishes.

Many people understand the reality of our situation (that no single act or object can fix the climate) while continuing to behave as though their act or object can actually do so. Knowing the truth of the matter does not diminish the social effects of the behavior carried out in the pursuit of the belief. In other words: we know, deeply and correctly, that technical objects cannot get us out of this mess, and we act, rightly and reasonably, as though they can. This is the logic of the fetish.

The significance of fetishes, like magic or wrestling, is not that people believe in them or don't. What matters are the dialogues of behavior and belief that fetishes gather and hold together—as well as the possibilities for social order that they facilitate. It is for these reasons that building better fetishes appears to be a worthwhile pathway for progressive reform in relation to climate crisis.

Fetishes can be understood as objects that both encompass the values of a society and function as channels through which social obligations and social orders are established and negotiated. Additionally, fetish objects are made of "heterogeneous components appropriated into an identity," albeit an identity that is composed of more than material elements.[2]

What are the heterogeneous elements gathered together by the ecological record? Until now, our discussion has focused on the materials, supply chains, and social relations that will be scaled up or down among companies reinventing what records are made of and how they are produced: PVC versus PET and PLA, injection versus compression, and so on. But it is equally necessary to consider the cultural desires, values, and commitments that such fetishes condense into objects and abstract into objectification.

The most prominent desires and values that inhabit records, as well as contemporary listening culture more generally, are extensions of a particular settlement on what an individual is, and what individuals want, in the context of the relationship between capitalism and consumerism that has been snowballing since the nineteenth century. It is no accident that the development of some of the greatest icons of industry and consumption—assembly lines in motordom, disassembly lines in meatpacking—coincide with the rise of the commercial recording industry and its culture of listening. Like the automobile industry and its car cultures, as well as the beef industry and its red meat republics, the history of recorded music has been built alongside a culture of mass production and individualized consumption. When people ask after the greenest way of listening to music, it is implicitly this set of values and expectations that they want to sustain (or overturn).[3]

This version of music is not intrinsically bad or good. Nor is it inevitable or permanent. Just as there are other ways of getting around besides driving a car and other diets besides meat-eating—and other things to value about mobility and cuisine—so are there other ways of listening. If the old saying about the coincidence of eating and being is true, this is because it is simply one expression of a deeper reality in human history. We are what we make.

It is worth drawing one more automotive parallel. In the world of wheels, lithium-ion batteries are touted as the power source of the future. Yet this type of climate solution only makes sense as a response to a certain conception of the climate problem. As a group of scalability critics asks, "Why is the scaling-up of electric vehicles seen as a more plausible response to climate change than the redesign of urban spaces around less car-centric forms of mobility?" If the distinction here is between car culture and public transportation, maybe the musical analogy is between record culture and public broadcasting. Or a public library.[4]

Radio and lending would certainly condense and abstract, in their objects and objectifications, as well as their scalable and non-scalable relationships, different forms of attachment than would records (ecological or otherwise). These possibilities are worth pursuing. At the same time, it may not be possible to retrofit the industrial and cultural formation of listening that has sedimented over the past 150 years. Or we may decide that, even if we could, we would not want to. We may choose to value individualized listening and mass access to recorded music. Even in such a formation, other kinds of digital listening are possible.

Around 2022, for example, a startup called Serenade introduced a format called digital pressing. The idea was to bring back the senses of fandom and emotional attachment to collectible objects that define formats of traditional tangibility, albeit in the form of digital-only album releases. Serenade did this by "pressing" a limited number of digital records, which means they stamped a computer file with a one-of-a-kind certificate of ownership and authenticity. The digital pressing, essentially a certified digital asset, would be stored on a decentralized peer-to-peer computer network. This network was accessed via Serenade's online marketplace, where the music could be heard. The network also kept an account of who owned the pressing. If the pressing was traded, they tracked the process and provided royalties in perpetuity to owners and copyright holders. Serenade intended digital pressing to be just like buying a record—without the record.

If this sounds like blockchain ledgers and non-fungible tokens, there is a reason for that. Serenade was a Web 3.0 marketplace and digital pressings were NFTs. Given the media hostility that instantly surrounded NFTs in the early 2020s, which disparaged not only their general absurdity but also their wasteful environmental impact (blockchain was said to use more energy than entire countries), it may seem odd to raise any of this in a discussion of music's ecological record. However, due to the specific way that digital pressings were created and maintained, they were less energy-intensive than the worst excesses of the blockchain system, especially of certain cryptocurrencies. At one time Serenade claimed that, in carbon terms, for every single vinyl LP made by the usual method they could produce almost 200,000 NFTs. Nick Jackman, previously Serenade's growth manager, and Mike Walsh, their former UK head of strategic partnerships, said that changes to their blockchain network, in September 2022, meant the ratio became more like 1 million digital pressings for every single LP.

The environmental question surrounding NFTs is not the only one. Criticisms come from all sides: assetization, financialization, de-democratization, taxation, fraudulence, the ongoing devaluation of musical work, and more. NFTs also operate nakedly in the service of capital. At the same time, some musicians have found that NFTs offer worthwhile forms for creativity and community as well as the possibility of better compensation for artists. In that context, if a format like digital pressing is a fetish that objectifies values and satisfies desires similar to those of "physical" formats, and if they do so at one-millionth of the emissions, maybe they shouldn't be written off just yet.[5]

Regardless, the NFT market has collapsed. By 2025, Serenade had abandoned digital pressings for a new product line called smart formats. Basically, the idea is that fans buy band merchandise that is embedded with near field communication chips. Pins, stickers, key rings—stuff like that. When tapped, the collectibles provide access to music on the Serenade platform. On the consumer end, the experience is similar to contactless payment on a smartphone, bank card, or turnstile. On the back end, it is not clear whether smart formats rely on the company's earlier technical infrastructure of blockchain and NFTs. Serenade declined to comment. Either way, the environmental angle to their marketing has become less prominent. But it is still there, down in their website FAQs, where smart formats are referred to as "content rich, but carbon light." Because chipped merch serves as the gateway to Serenade's digital platform, replacing "physical" media like LPs, smart formats are pitched to musicians as "a more sustainable way of supporting your work."

Although the NFT frenzy has died down, a new digital media mania has taken its place—and this development seems easier to write off, even if it is impossible to avoid. Artificial intelligence is marketed as a miraculous, unstoppable replacement for tasks that used to require sapient effort and discernment, like driving cars, detecting diseases, penning letters, and writing songs. AI can do some remarkable things, but it is still made of people. Computers can no more replace cognition than apples can replace oranges. It is a false equivalence.

In reality, *AI* is a term of art that references a jumble of technologies that are neither artificial nor intelligent and that, in their most hyped "generative" forms, are best thought of as synthetic content extrusion machines—Play-Doh Fun Factories pumping out digital lookalikes, soundalikes, and wannabes. Some musicians might get defensive here, if the history of popular music is anything to go by. They may insist that a generally critical view of AI misses the fact that some listeners have a rich and varied appreciation of artificially intelligent songs and playlists, some acts do interesting things with AI tools, and some synthetically extruded musical content has merit. Nothing in the general critique denies that AI-generated music can be enjoyable, creative, or valuable. It's simply that no amount of positive experience with the technology, no amount of genre-bending use cases, no amount of poptimism can neutralize the more basic social mediations of AI as a site of political contest.[6]

On top of the blinkers and biases of the algorithms built into these machines, AI has quickly been linked to discouraging statistics about carbon and resources. Extrusion means extraction. It means exploitation. In a moment when synthetic songs are topping the Billboard charts and AI music executives are dreaming of spamming the world with billions of automated tracks every year—Spotify's mood music made by machines for other machines, Suno's novelty tunes for our Valentines and friends named Earl—how many of these trinkets and trifles are destined for the junk drawer of history? Is this a marvelous new era of music, or simply the latest chapter in the impoverished and unsustainable chronicles of crap?[7]

As the possibilities and limitations of NFTs, Web 3.0, and AI continue to unfold, the fact remains that the most common form of listening today is streaming, and this version of musical industry and culture is the most resource-intensive listening formation the world has known. This is due neither to the technology itself nor to the desires of individual listeners, but to the forms of economy and scale that streaming serves.

There is a growing number of responses to these issues. One is subvert.fm, an online platform and marketplace modeled on cooperative ownership with a fifty-year plan to become the Mondragon of music. Another response is vilvit.io. Vilvit, pronounced like *velvet*, also wants to change the infrastructure of streaming. Its founders call for cooperation among streaming services and digital service providers. They note how streaming is based on old plumbing that was built on competition. This created substantial inefficiencies and redundancies across the record industry. "Today," Vilvit says, "every streaming service hosts almost the same catalog as the other. Over and over again. Petabytes duplicated." Unless something changes, the situation will only get worse.

Vilvit's proposal for change is inspired by the Svalbard Global Seed Vault, a biorepository for the world's crops. Instead of seeds, though, Vilvit wants to create a centralized bank of all the world's recorded music—one that could be used by streaming companies and musicians everywhere. In addition to eliminating vast and irrationally redoubled efforts in uploading, archiving, and distributing musical data, which itself would save plenty of energy, the Vilvit initiative would have the added benefit of being powered by water and wind, and its computer servers would stay cool by virtue of being tucked in a Norwegian mountainside.

When I spoke to Vilvit, the initiative was in what they described as a drawing-board phase. Later, they vanished from the internet. The idea is to regroup and refocus. To spend less time talking about the mission, more time doing it. In any case, the issues raised by vilvit.io are not new. According to industry insiders, including chief technology officers at the major music labels, the inefficiencies of streaming have always been clear, and the issues were brought to executive attention in the early 2000s. But the CTOs' concerns were dismissed, because addressing the distributed technical inefficiencies of streaming would have been too expensive. Now, with the situation of the climate and the rise of green capital, such issues are not only unavoidable.

They could be profitable. Twenty years on, the executives may finally be listening.

Ecological records, radios, libraries, non-fungible tokens, artificial intelligence, digital vaults—all point to possibilities for doing things differently; some might help build better fetishes. None of them is going to solve the climate crisis on its own, not even in the relatively small realm of music. But since technical solutions are such a major part of music's recomposition—and since knowing the limitations of our beliefs does not stop them from having practical effects—the goal here has been to provide a particular political vision of what records are, how they come to be, and what they might do. It has been to take equal stock of how we make ourselves and how we relate to the world around us, not exactly through our objects but rather through the objectifications that objects condense and regularize, through the social mediation of the fetish.

Fetishes are always and necessarily forms of encompassing others and objectifying relationships. It so happens that, under our particular economic arrangement, fetishes take on a commodity form that tends toward relations of domination and exploitation, exteriority and alterity. Although it is not possible to completely escape any of this, as long as capitalism predominates, it is nevertheless possible to try producing objects that establish, as best they can, different relations of commitment, obligation, accountability, and maybe even reciprocity.

Going back to Harvey Molotch, the goal is first to understand these processes so that we may produce more socially and ecologically responsible goods. It is then to know that producing better goods, or fetishes, is an avenue for instantiating human development and initiating social progress, however limited the gains may be under capital.

The fetish was born of encounter. Although the term *fetish* originated in the medieval church, in connection with icons of witchcraft, its modern meaning comes down from the fifteenth century. It was then that European merchants imposed particular relations of exchange on cultures in West Africa. Through the Portuguese, and later the Dutch, fetishism came to describe unfamiliar practices by which unfamiliar African cultures seemed to turn any old thing into a god—to sacralize trivial objects and imbue them with divine powers to bring rain or cause drought, and so on. Applying the term *fetish* to such practices was a form of condescension, an accusation of superstition meant to imply not only that "African peoples were too immature to perceive the world correctly" but also that "materially embodied African gods" were "the universal counterexample of proper reasoning, commerce, governance, and sexuality."[8]

Ironically, while this understanding of fetishism was never a good description of what was going on in these African cultures, it was very revealing as a mirror of certain European tendencies, values, and perceptions of the world—not because they were so utterly different from the ones observed in West Africa but instead because they were essentially the same and, as such, uncomfortably familiar. In this way, the fetish was, according to anthropologists, "premised on the invocation of a surpassed primitivity" even as "the concept was sutured into an Enlightenment project for which it functioned as the signifier of a constitutive alterity."[9]

If the history of the fetish is a history of encounter, then it is a violent, colonial encounter. For this reason, we might question why it is useful or even possible today to continue employing an idea that is so inseparable from the history of racism. In other words, "why risk," asks researcher Rosalind Morris, "reviving the ghosts of an epistemic violence whose victims, in Africa and elsewhere in the not-yet-decolonized and now-recolonizing world, are once again being mobilized to supply new forms of capitalism with its medium and material of expansion?"[10]

One answer is that the concept and phenomenon's value obtains precisely in its unavoidable invocation of these past and present realities. The fetish sustains evidence of ongoing encounters with (and encompassments of) selves and others, people and things, in material culture. It serves as a constant reminder that the condensations of objects and the abstractions of objectification may strive to open shared spaces for political contestation and the perception of subjection. Or they may shut down such spaces. To deny these encounters and encompassments is to deny the realities of others.

Yet this is precisely what many technical solutions to climate crisis do, in music as elsewhere. Insofar as technical solutions take emissions as their main focus, urgent as such reductions are, they also lock in existing forms of exploitation and oppression—impacting not only on the direct subjects of capital but also various other social categories that likewise arise in the production of surplus value. Focusing on emissions reductions primarily at the points of distribution and consumption in the economic core—which is to overlook that any technical climate solution will necessarily encompass many peripheral people in many peripheral places who have to work the earth, process materials, and make technologies—inflates the problems and solutions of a global middle class into a universal context.

All this malevolent encompassment will continue so long as climate solutionism places an especially strong emphasis on one particular dimension of the challenge we face. Carbon.

Part II
Institutional Solutions

7
The Carbon Question

Thirty-seven tenths of a gram. The weight of a pea. According to the telltale gauge at the bottom of the web page, that is how much carbon dioxide was exhaled by a single click of the mouse in early 2023. It was a cleaner click than 60 percent of other websites in that moment, and its underlying electricity requirement was nothing compared to the 400 terawatt-hours used annually by the internet. To put things in perspective, 400 terawatt-hours is about one-tenth the annual consumption of the United States. It is more electricity than all of Britain uses in a year. That's a lot of peas.[1]

I generated this fractional breath of CO_2 when checking on ClimateEQ, a music sector–specific carbon reduction and training consultancy that was "launched in 2021 to help the music industry work towards a low carbon future" and whose clients include major labels like Sony and Universal, independents like Warp, and brokers like Key Production. A year before my follow-up click, ClimateEQ had delivered the remote course that qualified me as someone who has "met all the requirements of the Carbon Literacy Standard and thus for the purposes of workplace, education, and community should be regarded as Carbon Literate." So says my certificate.

There are nearly 150,000 of us "carbon literate citizens" across more than 10,000 organizations. The numbers are rapidly growing, and they reflect the reach of the Carbon Literacy Project, which is the Britain-based body that accredits programs like ClimateEQ, among many others.

The Carbon Literacy Project defines their object as "an awareness of the carbon costs and impacts of everyday activities." Their mission is to provide "relevant climate change learning" in a way that "catalyzes action to reduce greenhouse gas emissions." As a certified carbon literate citizen, I am supposed to possess the knowledge and skills to understand and abate both my own carbon footprint as well as the footprints of others. I am also supposed to be more employable.[2]

If these are the markers of carbon literacy and carbon citizenship, it was on the third and final day of our ClimateEQ training that my classmates and I learned the first and foremost priority in the "climate strategy hierarchy": measurement. Above all else, responsible carbon literate citizens must "understand and report your impacts and track how you're doing year by year."

Sitting in a drafty Montreal apartment, as the overnight forecast threatened forty below with the windchill, my mind drifted from the task at hand. The carbon literacy training is presented in a distinct language of precision, progress, positivity—and the benefits of education, education, education. But didn't sociologists say institutions and credentials sometimes support the same old, same old? Hadn't historians pointed out that the meaning of measurement is not self-evident? Weren't critics warning that focusing too closely on carbon could obstruct a wider view of ecological devastation and human suffering? And what exactly was carbon, anyway?

I muted my mic, turned off my camera, cranked up the heat, and wondered.

8
ClimateEQ

When James Dove founded ClimateEQ in 2021, he was dealing with a self-diagnosed case of climate grief. The symptoms had been with him for years, effects of prolonged exposure to the news. The more he paid attention, the more the severity of the situation upset him. How was it possible that everyone was walking around like nothing was happening? He needed to do—something.

Having spent much of his life playing music and working in the industry, it was unclear to Dove what path he might pursue, not just to alleviate his grief but also to do some good in the world. "I'm not a scientist. I'm not a CEO. I'm not a politician," Dove says. "What difference can I actually make?"

To find out, he started doing his own research. Mostly online, at first. Eventually Dove completed a number of climate literacy courses like the one he would later launch. What he learned in those courses brought him to a crossroad. Should he change careers and go into environmental science? Perhaps not. As Dove had discovered, the science was clear. It was the communication of the science, he thought, that needed a boost. Dove had experience in music communication and connections in the music industry, plus the Carbon

ClimateEQ priorities. Image by ClimateEQ.

Literacy Project did not have a music-sector training program. Once those stars aligned, it was not long before Dove's own training course was accredited and accepting students.

The principles of ClimateEQ, Dove explains, can be summarized in three points. First, carbon reduction is better than carbon offsetting. Second, climate communication should balance "the right amounts of severity with the right amounts of solutions," and it should be based in the latest science, so that people can make informed decisions. Third, climate crisis is traumatic, so "it's important to sit with your feelings"—to accept them, to move with them, to manage them. Because "emotions drive people" and "people drive change."

These principles constitute the triple meaning of the EQ in ClimateEQ. The philosophy of equalization (EQ) in music is about subtracting frequencies more than adding them. The idea of balance in communication is about establishing an equilibrium (EQ) between harsh realities and hopeful possibilities. And the notion of managing feelings requires a certain emotional intelligence quotient (EQ, after IQ).

Later in the same year I was carbon-certified, ClimateEQ had already revised its training course. It expanded beyond music to the creative industries more generally, as well as entertainment, sport, and small businesses of all sorts. The expansion coincided with the arrival of Anthony Daly as ClimateEQ's codirector. Daly brings a long career in carbon, especially with regard to measurement, quota setting, and reductions initiatives in the supply chains of the music and entertainment industries.

Dove and Daly have different backgrounds. They have different skill sets. But, says Daly, they both want to do the right thing when it comes to the climate crisis. "Let's get people on board and let's get reducing," he says. In the service of that goal, Dove and Daly have joined forces. They are a bit like Batman and Robin, Dove adds.

The basic contours of James Dove's story conform to a broad genre of climate enlightenment autobiography. The genre begins in helplessness and despair, because nothing or not enough is being done to fight the evil that each day is visited on the earth. Then there is a period of education, both autodidactic and formal. Finally, protagonists emerge with something of their own to offer the world—something that just might save it.

There are similarities here to the plots of comic books and the archetypes of superheroes. Far from making light of climate autobiography or its protagonists, this suggests that the genre and its characters must be taken seriously. The genre expresses something about a social reality wherein people take it on themselves to manage things that were once understood as the duties of a collective authority, like a government.

In other words, distributed expectations of political accountability have devolved into concentrated senses of personal accountability. Social problems are increasingly individual problems. The term for this process is *responsibilization*.[1]

There are things to be said for being responsible—for taking ownership of our relationships to the climate, either as individuals (sympathetic consumers) or entrepreneurs (sympathetic producers). Yet responsibilization is about more than ethical orientations or codes of conduct. Responsibilization reshapes how people think of themselves and how they exist in the world, both in relation to one another and in relation to their institutions. This is explicit in the carbon literacy training, where the goal is to produce "carbon literate citizens"—150,000 and counting.[2]

The actual number is much higher. What is handed down from the accrediting body and presented in the literacy training is in many ways a certified version of the climate common sense, or mainstream environmental thinking, that saturates the media. Dove and Daly recognize this. They know that carbon is only part of the picture, and they advocate for larger systemic change. Yet in both the cases of common sense and the wider Carbon Literacy Project, the first move is always to take responsibility—to educate ourselves as private individuals (consumers, employees, entrepreneurs) in order to understand our own footprints so that we may walk the walk.

Responsibilization comes with an implicit theory of change, an implicit structure of political engagement. This is known as incrementalism. Here the main struggle is not directed outward, against institutions, economic arrangements, or social architectures. Instead, the starting point for struggle is focused inward. Individual feelings of climate guilt, shame, and helplessness are diagnosed. Personal remedies are proposed and pursued. Then there are regimes of self-improvement, based in a combination of scientific and moral education. There are forms of lifestyle politics that revolve around buying the right products and starting the right businesses—or boycotting them. The assumption is that all these small, individual steps are the start of something bigger.

Unfortunately, this 1+1+1 theory of change does not have much corroborating evidence. This is especially true in terms of environmentalism, where, according to researcher Samantha MacBride, turning to "the self, the local, the doable, the manageable, the small step" can be less a strategy of resistance than one of "resignation and powerlessness self-cloaked in optimism and reassurance." Incrementalism, certainly in terms of carbon, tends to undercut itself. In the guise of climate progress, it secures the status quo. It is a better explanation for the continuity of the social order than the disruption of that order.[3]

Measurement is the foundation of responsible carbon literacy and carbon citizenship. It is a fundamental part of the public, economic, and political common sense surrounding climate crisis, fixated as this common sense is on weighing carbon accurately, calibrating its scales precisely, and standardizing these procedures widely. The underlying philosophy is that measuring is knowing and that knowing means changing. Such is the hope of carbon metrology.

If it seems like nothing could be more natural than measurement, it is worth remembering a classic critical dictum. When things feel most natural, this is sometimes when ideology is doing its strongest work to conceal the historical and political character of a given situation. This is the case with numerous social phenomena, like gender relations or family structures. It can also be the case with physical phenomena, like the boiling point of water. That water boils at 100 degrees appears to be given in nature, but actually this is a misconception, a settlement of convenience and convention. (The boiling point of water varies according to factors including not only elevation but also the container being used to heat the water, how gently or aggressively the water is heated, and many others.) Other settlements are possible.[4]

As in thermometry, so in carbon measurement. To understand the meaning of this, it is helpful to return to a moment in the history of precision that lies at the base of nearly all official measurements today. That moment is the invention of the metric system in eighteenth-century France.

During the 1700s, to engage in French trade was to encounter a quarter-million different units of measurement. These units, with their hundreds of systems, were based in anthropometric references, both in terms of human anatomies and local topographies. There were feet of this and thumbs of that, certainly. There were also lengths of cloth defined by the width of a regional loom, and weights of coal defined as one-twelfth of what a miner could dig in a day. By the mid to late 1800s, France—and much of the world—had done away with these messy, local, human referents. In their place stood a single, unified metric system based on universal constants in the natural world. What happened?

A textbook explanation might describe the invention and institution of metric measurement in terms of newfound precision and objectivity, a major advance in scientific progress. There is no doubt that the metric system matters in those terms. However, a fuller description requires understanding metrification as an agent of the French Revolution.

During the revolution, as one group of elites (the would-be bourgeoisie) worked to abolish the feudal and dynastic orders of another (the nobility and the monarchy), a novel system of measures was consciously introduced by the revolutionaries, not to advance science but rather "to mediate a fundamental tension between state authority and the construction of a market economy." In other words, this new measurement system was "deliberately crafted to break the hold of the Old Regime's political economy" and to "ensure the triumph of free trade." Standard units of measurement made the anthropometric cacophony commensurable (and controllable) from town to town, across the land, and eventually around the globe. It made quantities newly available to an emerging sphere of exchange. The metric system was created for a particular market system.[5]

Local French communities hated the metric reformation. For them, it represented both a dramatic upset in their daily social order and a relatively bland shift from one form of subjection (to the nobility) to another (the bourgeoisie). For the revolutionaries themselves, though, the benefits were clear—and they extended beyond trade. Metrification helped establish a generally free market certainly, but it also crafted a specifically useful French population. Metrification embodied a form of rationality that, absorbed by the people, would help turn them into a rational citizenry—subjects whose thoughts and abilities and concerns would be aligned with their role in the market. In short, the revolutionaries "saw the metric reform as a crucial stage in the education of modern *Homo economicus*."[6]

Carbon metrology accomplishes something similar. Although the language today is less about rationality than responsibility, there is a kindred drive to establish calculative agencies, toward precise atmospheric accounting, toward drawing various knowable and exchangeable equivalencies (whether pea weights per website click or offset dollars per carbon kilo). At this level, the purpose of standardized carbon literacy, just like carbon accounting more generally, is to make emissions available to the arenas of responsibilization and exchange.

As with the metric revolution, there is today with carbon a similarly explicit belief about the connection between how we measure and who we are. This includes forms of self-discipline and even self-deprivation that, although framed in the commonsense terms of doing one's part to save the planet, and while surely important in terms of establishing ethical orientations to the world, are foundational to creating a population that not only accepts but also welcomes and actively institutes the shifting regime of accumulation.

Responsible carbon literacy makes for a responsible carbon citizenry. The metrology of carbon, in music and beyond, may be a crucial stage in the evolution of a new figuration of what it is to be human. *Homo carbonicus*.

9
Atmospheric Accounting

Some of music's most exact measurements have been taken in relation to concerts. Spend any time looking at music from an environmental perspective and you are sure to learn that live music accounts for the majority of the industry's carbon footprint. You will also quickly learn that most of that footprint comes from audience travel.

According to the latest and most comprehensive study ever undertaken on the subject, gigs generate over 14 million tons of emissions annually in the United States, accounting for 0.2 percent of overall greenhouse gases. In Britain, that number is more like 4 million tons and is good for 1 percent of total yearly emissions. Getting to those gigs accounts for over 60 percent of live music's footprint in the United States and nearly 80 percent in Britain. Discovering that audience travel is "the single largest source" of music industry emissions is the report's "key insight."[1]

This insight may be key, but it is not new. Singling out live music and audience travel in accounts of music's carbon footprint has been industry orthodoxy for at least two decades—certainly since 2007, when both Radiohead and the arts-and-climate nonprofit Julie's Bicycle published carbon audits of the music world. Radiohead's efforts were focused on the band's North American concerts, finding that "fan travel and consumption make up 86 percent of the theater tour and 97 percent of the amphitheater tour." Julie's Bicycle, meanwhile, in a sector-wide study that included recording, publishing, and more found that "live music performance together with audience travel account for three-quarters of the UK music industry's emissions."[2]

Musicians are not blameless here. Some touring acts, like Taylor Swift, catch flak for their private jet emissions and try to compensate through carbon offsetting. Others, like Massive Attack, pause or downscale their touring activities while seeking low-carbon options. "Artists should be held accountable for their travel," according to interviewees in the recent study. And artists apparently agree. "Still," say the interviewees, "their emissions are minutia when compared to that of fans." It is only logical that if "fan travel is the largest source of GHG emissions in the live music industry," then "efforts for reducing emissions should be largely directed there." The report aims to "nullify the distracting, ill-informed mischaracterization of the emissions impact of artists" and make audiences responsible for developing "good habits" as they attend concerts. Admit one: *Homo carbonicus*.

None of these environmental reports looks exclusively at audiences. They all mention the impact of wider business practices like food and freight. They recognize the tangled and systemic character of the issues they face. But the category of audience travel comes up repeatedly in studies of live music's carbon footprint—and it is never really questioned. The category is given. The only task is to measure it with greater precision and more impressive methods, so that the largest carbon reductions may be achieved. There will be successes on that front, even as its underlying beliefs about measurement and responsibility lift some eyebrows. But fixating on carbon in this way carries an unintended consequence. It preserves a false dichotomy between artists and audiences.

Performers on stages and attendees in venues are obviously engaged in different kinds of activity. But this surface difference papers over something more fundamentally shared. Most musicians and fans need jobs to make a living. Even very famous musicians are often contracted to sell their capacity to work. Although the carbon footprints of artists and audiences can obviously be measured separately and reckoned accordingly, both groups have to burn carbon in order to participate in social life. And most of them did not choose the economic arrangement that compels them to burn that carbon, that splits labor and leisure in this way, that organizes work and pleasure around certain kinds of mobility.

This setup sometimes authorizes cheap shots about the supposed irony that environmentally conscious musicians and fans would travel for concerts. It can permit low blows about events that promote environmentalism while also making claims on the environment. But does it count as insight to point out that people happen to be part of the reality they want to change? Is there really no festival without cruelty? Either way, environmental thinking on live music encourages a form of atmospheric accounting where scrutiny of the personal trees passes for consideration of the political forest. Focusing on audience travel is one more way that economic power disorganizes and discourages its potential opposition.

Homo carbonicus does not actually exist. It cannot be mapped onto any individual's thoughts or actions or admitted to any concert. Rather, *Homo carbonicus* is a figurative abstraction, a collective consciousness, an interpersonal framework for defining problems and seeking solutions. If the go-to responses of this figure are responsibility and metrology, such instincts find stability in institutions that objectify certain calculative agencies in a wider audit culture. In other words, *Homo carbonicus* is part of a more general historical moment in which people have been transformed "into 'auditable' entities that focus their energies on doing 'what counts' rather than what is necessarily moral or right."[3]

These intuitions and institutions can operate differently depending on their scale and size, but they share some patterns of thought. One of the most important limitations here is that in such settings, instead of asking what counts, we assume we already know, and we focus instead on how to account. This explains the focus on audience travel in the carbon footprint of live music. It also helps explain the drive for ever more precise carbon calculators and ever more detailed atmospheric accounting in other industrial and cultural areas of music, while larger questions about carbon go unasked.

Custom carbon calculators are proliferating among independent music businesses and organizations. These range from free personal tools, like Onboard:Earth (a smartphone app that measures and tracks carbon emissions from musician and fan travel to concerts) to Green Your Noise (an online service that will estimate the emissions of everything from single concerts to larger tours and productions). Others follow membership and subscription models, both paid and not. Some hire consulting firms. One of the most visible tools here is the Independent Music Companies Association IMPALA Carbon Calculator, developed in collaboration with the nonprofit climate organization Julie's Bicycle.

The IMPALA Carbon Calculator is tailored to independent record labels. It is designed to measure business activities deemed to be within IMPALA's "operational boundary."

Such activities include office use, employee commuting, business travel, manufacturing, and distribution. Since launching in 2022, IMPALA's calculator has become the tool of choice for 100 labels in twenty-one countries.[4]

Music's largest companies do things differently. Since around 2020, the three major music groups, Sony, Universal, and Warner, along with large streaming providers like Spotify, have all started reporting their emissions figures annually. An important starting point in the process is making splashy appointments in corporate social responsibility and environmental social governance. Warner, for example, hired Samantha Sims as head of ESG in 2021. Sims worked previously to establish a sustainability program at PVH, which owns clothing brands including Calvin Klein and Tommy Hilfiger. Spotify hired Elizabeth Nieto in 2021 as leader of equity and impact, an umbrella role that covers initiatives in sustainability, social impact, and "early career pipeline" as well as "diversity, inclusion, and belonging." Nieto's prior jobs were with corporations like Citigroup, MetLife, and Amazon.[5]

Despite differences between the approaches of the independents and the majors, the back-end work of calculating carbon is essentially the same. To inventory their emissions, carbon's bookkeepers first consult the Greenhouse Gas Protocol Corporate Accounting and Reporting Standard. The GHG Protocol is a process document. It helps businesses decide how and where to draw the lines of responsibility regarding emissions. These are the "operational boundaries" mentioned by IMPALA.

Once those boundaries have been mapped and tallied—counting how many kilometers an organization's employees have traveled through the air, how much electricity the offices have used, and so on—the second step of carbon accounting involves applying an emissions conversion factor in order to translate those figures into greenhouse gases. If your business is based in Britain, the Department for Business, Energy, and Industrial Strategy will tell you that taking a domestic flight emits 0.24587 kilograms of carbon per passenger, per kilometer, or that staying in a hotel emits 10.4 kilograms of carbon per room, per night. If a company is based primarily in the United States, the conversion factors are likely to be different—and probably provided by the Environmental Protection Agency. A common, worldwide source for power usage is the International Energy Agency's annual emissions factors reporting.

The GHG Protocol is known to be accurate when it comes to understanding the emissions from an organization's onsite activities as well as the energy it purchases from the electricity grid. Yet the protocol is open to interpretation when it comes to accounting for emissions along an organization's wider supply chains and waste streams. Still, it is less important in this context to question accuracy than the boundaries of responsibility.

For example, the GHG Protocol is what allows Spotify to stress that 99 percent of their emissions fall outside of their direct control. It is also what allows IMPALA to say that, although independent labels are answerable for the manufacturing emissions associated with their products (like records), "Our conclusion is that labels are not responsible for digital services' emissions, in the same way that we are not responsible for the emissions of physical retailers' stores on the high street." This is a curious move, where an organization's wider emissions are both ushered into the accounting picture and bounced out of it. If responsibilization in one sense means accepting forms of self-monitoring and self-deprivation, taking responsibility in carbon accounting can also be an evasive maneuver.[6]

This is not surprising, given the history of the GHG Protocol. Although national carbon accounting and reporting goes back to 1995, with the United Nations Framework Convention on Climate Change and the Kyoto Protocol, and while individual carbon footprints entered climate common sense only after 2004, when the idea was popularized by an ad campaign spearheaded by British Petroleum and the Ogilvy marketing agency, corporate carbon measurements and methods took shape in 1997. It was then that BP, wishing to reduce its emissions profile, found there was no standard system for measuring and reporting on carbon at the corporate level. When BP set out to develop such a standard, its efforts were noticed by Monsanto, General Motors, and the World Resources Institute (currently funded by the Jeff Bezos Earth Fund, the World Economic Forum, Walmart, and many others). The collaboration resulted in a 1998 report titled *Safe Climate, Sound Business*. This report, alongside additional collaboration with the World Business Council for Sustainable Development (which represents not only BP but also DuPont, Shell, and more), laid the groundwork for the GHG Protocol.

While such efforts seem good if taken at face value, the fact that they are driven by market considerations means there are ulterior purposes at play. For example, when BP and other corporations talk about reducing their emissions, the word *reduce* is doing some heavy lifting. Corporations are often not seeking direct or absolute reduction. They are seeking the appearance of reduction that comes from emissions trading schemes—which are controversial at best, ineffective at worst, and everywhere ensconced in the ecoliberal conviction that capitalist markets, purified of their imperfections and filtered of their failures, are the solutions to their own problems.

Another ulterior purpose of the GHG Protocol and its carbon calculus is that, while it helps corporations appear environmentally virtuous by way of self-scrutiny and self-regulation, the goal is in fact to anticipate (and therefore lessen the impacts of) more strenuous scrutiny and regulations that may be introduced by governments and international organizations. This is straight out of the corporate playbook. Act, or be told how to act. Get out ahead and set the model. Then lobby policymakers to shape laws around the model that the corporations helped set in the first place.

Political scientists have observed that, when it comes to global environmental governance, the GHG Protocol and related initiatives are not only about getting ahead of regulations. Private actors, such as corporations and entrepreneurs, are themselves becoming the regulators—or at least they are exercising such strong influence over the regulatory process that the private actors are effectively in charge. In a way that mirrors (but inverts) responsibilization at the individual level, private firms are assuming "duties normally considered the province of governments." Music's largest businesses may not be as active or influential in such processes when compared to a corporation like BP, but music is part of the same story.[7]

The GHG Protocol is a bible of atmospheric accounting. Despite some uncertainty around how it defines wider supply chain emissions, the authority of the protocol is not in doubt. Most of the questions and complexities surrounding atmospheric accounting are pursued during its second stage: applying emissions conversion factors. A lot of effort goes into those conversions.

Consider the collaboration between Terra Lumina Consulting, "an Austin-based, woman-owned consulting firm focused on equitable solutions to the climate crisis," and the Secretly Group, a family of record labels whose roster probably includes at least one of your favorite acts. Secretly's cofounder, Ben Swanson, along with others in the company, has been formally pursuing sustainability since 2021, publishing a detailed analysis of the company's carbon footprint on its website. Secretly was an early investor in the IMPALA Carbon Calculator, which they have used for some of its emissions inventories. However, the IMPALA calculator is based on UK emissions factors. To get more accurate figures for Secretly, which is headquartered in the United States, the conversions have to be done with US data.

Jen Cregar, founder of Terra Lumina, offers an example that illustrates the level of detail involved in such work—as well as just how quickly things can get complicated. Say she's calculating Secretly's business travel and freight distribution, both on the ground and in the air. Because US fuel blends differ from those in Britain, Cregar will have to seek out and then apply different conversion factors. Plus the US train system is different. The power grid is different. And, at the other end of a product's lifecycle, what about the landfill? What kind is it? Is the garbage burned in an incinerator? If so, how old is that incinerator? How far is it from the city? Add to these the manual work of locating various invoices, translating across billing cycles, correcting errors, and so on. Finding the conversion factors, pulling the data, cleaning it up—Cregar is sure that this will someday be a job for artificial intelligence. For now, it is a detailed, unending, manual effort.

In carbon calculus and atmospheric accounting, the complexities of conversion tend to infinity. It is in that space where people like Jen Cregar and companies like the Secretly Group are doing important, dedicated work. Yet it is also in this asymptotic space where climate common sense casts some of its longest shadows.

What if we could know carbon completely? There would be benefits, of course, notably in terms of strategy—in terms of where to focus efforts at emissions reduction or elimination. This is one reason that the work of atmospheric accounting is so important. It also explains something about its allure. The temptation is to keep pushing atmospheric accounting further in the direction of precision, deeper into total measurement. Yet simply doubling down on capital-driven protocols of accountability and reporting is unlikely to fix the flaws of an arrangement that feeds on those very flaws. The best way out of a hole is usually not to dig in.

More holistic frameworks for understanding the ecological stakes of capital exist. Environmental life cycle assessments of various kinds, for example, can be traced back to the 1960s. During this decade, companies like Coca-Cola began commissioning research to examine their supply chains and waste streams. They did so to compare different products and figure out which ones were "better" from an environmental standpoint. Although such work was initially directed internally, it has since taken hold of the imaginations of sympathetic producers and sympathetic consumers. It is part of what defines climate common sense.

Life cycle assessment cannot be subjected to all the same criticisms as atmospheric accounting. Although carbon does feature prominently in LCA, the framework is expressly concerned with wider questions of energy, resources, and waste as goods and services circulate in a globalized economy. Its prominence is rising in the music world.

GZ Media, the Czech company said to be the world's largest vinyl manufacturer, published in 2025 the "industry's first life cycle assessment of a vinyl record." Their analysis is not quite cradle to grave or cradle to cradle, which are the most ambitious types of LCA. GZ's assessment is cradle to gate, covering the part of the life cycle from when raw materials are pulled out of the ground to the moment a finished product leaves the factory. In this light, GZ's findings suggest that the most consequential reductions in making vinyl will come from using renewably sourced electricity during the pressing process.[8]

The Music Climate Pact, a Britain-based agreement for the music industry to attain net zero emissions by 2050—which includes all three major labels as well as numerous independents and other organizations, along with the Vinyl Alliance, an international trade association for the record industry—have collaborated on another life cycle account of the disc format. Their report aims "to transition the record manufacturing value chain to a sustainable model" by standardizing the formula for quantifying emissions and identifying hot spots for carbon reduction across the life cycles of both traditional records and injection-molded PET discs. The Music Climate Pact is also tackling the notoriously difficult task of collaborating with streaming platforms, which tend toward corporate secrecy, not least in terms of their environmental impact, to come up with accurate figures for emissions and reductions in the digital world. The results of this work are providing the foundation for some of the most concerted, detailed, and substantive climate actions in the music industry to date.[9]

There are proposals to maximize the remit of life cycle assessment, to bring the methodology into music research and product design. The added emphases here would be on social hardships as well as questions of goodness and beauty, on making the instruments of musical consumption and production in ways that would be "aesthetically and ethically humane" and that would "value the integrity and stability of the earth." These would be welcome developments. Similar lines of inquiry have been increasingly prominent in the music world over the past decade.[10]

All this work is crucial for the music industry's commitment to environmental action, but its limitations should be kept in full view. One issue is that LCA is necessarily caught in the same traps as atmospheric accounting and greenwashing. It is now (and to an extent has always been) part of the theater of self-scrutiny by which brands create value in times of ecological crisis.

It is also possible to criticize the analytical framework for the quality and completeness of its data, about its pretentions to objectivity. None of this can be perfectly achieved. Yet a myth of total measurement persists, suggesting that if only we could measure and represent the ecological reality of capital fully and completely we would inevitably be driven to lessen its effects.

LCA may be more detailed and more socially aware than other forms of ecological accounting. It may even help us produce kinder and more beautiful things. But LCA still tends to fold solutions back into the problems they try to address. The underlying issue is that the relationship between knowing the "is" and making the "ought" is not as straightforward as we might hope.

10
Future Energy Artists

From Lincoln's Rock, overlooking the Jamison Valley in Australia's first state, you can see a shortwave haze of livid light. It blankets the canopy as it rolls out over the horizon. The scene is rendered by sunbeams scattering off oil droplets released by the far-reaching eucalypt forests below, an atmospheric effect so prominent that it even tinted the smoke rising from the region's fires during the Black Summer of 2019–2020. Such is the phenomenon that gives this place its name. The Blue Mountains.

The bush—this bush—is her mother, Heidi Lenffer says, walking outside her home west of Sydney, not far from where she grew up. It seems fitting that we have met here, in this part of the country, to discuss Lenffer's work as a musician turned environmentalist-entrepreneur. The area has suffered unheard-of losses due to bushfires since the early 2010s, and mournful images of the Blue Mountains burning were among the inescapable symbols of climate crisis and political inaction in midst of that Black Summer. Fire took Lenffer's own home in 2013, which stood previously on the site of our meeting.[1]

The Jamison Valley in New South Wales, Australia, 2022.

Witnessing the rebuild of that home instilled in Lenffer something of an operational philosophy regarding the state of the climate. The job required looking back and sifting through debris but also forging ahead, creating something new. Grieving plus gumption. Likewise, with the climate, Lenffer mourns the things we have lost in the fire. But she lives in hope—and action.

Further environmental stirrings came later, in ways that parallel the climate enlightenment autobiography of James Dove and many others. After years of touring with her band Cloud Control, Lenffer understood that to be a professional musician is also to be a professional traveler. Which is to be a professional carbon emitter. Uncomfortable as she was with that reality, the go-to solution at the time, carbon offsetting, was known to be limited. Offsetting did not match the magnitude of her vision. This was around 2017.

By 2019 Lenffer had launched Future Energy Artists, or FEAT, an investment platform geared toward musicians, allowing them to help finance solar energy projects via investing either lump sums or touring profit percentages. By 2021, having invested its nest egg in a managed fund that provided financing to a construction company, FEAT oversaw the development of a fully operational solar farm.

The Brigalow solar farm is located on a droughted grain field in rural Queensland, nearly 300 kilometers northwest of Brisbane. Brigalow covers 80 hectares (about 150 football fields) and generates thirty-five megawatts of electricity (enough to power over 10,000 homes), which it will do throughout its intended life span of three decades. The photovoltaic farm generates something else as well. Returns on investment. It is a success, a landmark project, the first music industry–backed solar installation.[2]

There is no single factor that led Heidi Lenffer to investment finance and solar cells. On her first visit to the Brigalow site, encountering terrain so parched and cracked that the region's snakes had migrated from the surface into the fissures, Lenffer describes making "small dust explosions kicking the earth with steel-capped boots" as she reflected on "the long and nonlinear journey of brainstorms, artist chats, calls with scientists, emails, partnership convos, and Attenborough docos that brought me to that patch of dirt."

One path Lenffer considered was establishing a climate charity or nonprofit. But, like offsetting, the idea did not fit her aspirations. If you put a dollar into a charity, she says, you trust it will do some good, but essentially the money is gone. If you put money into an investment scheme, on the other hand, there is a rope tying you to that dollar. It is a relationship. Additionally, as a musician Lenffer knew that most bands are financially insecure and that donating their time for, say, benefit concerts can actually be a double cost to the performers. In bringing a cause-oriented project to the music world, the former Cloud Control keyboardist did not want to replicate those patterns. Instead, she wanted something that would give back to musicians and incentivize them toward ongoing participation in the climate movement. Investment ticked those boxes.

Part of the appeal of solar, Lenffer says, is that it looks and feels as momentous as the problem we face. Solar farms are gigantic, tangible, infrastructural, and awesome. They reflect the ambition and the name of Lenffer's initiative: FEAT. Solar also represented a real and present yet underused alternative to fossil fuels in Australia, where 70 percent of electricity comes from coal and gas. Plus, Lenffer adds with defiance in her voice, in those cases where Australian solar projects are being developed, it tends to be large international banks and energy corporations coming in and buying up land. The rich are getting richer. She asks: "Why can't local musicians play a role in the energy transition?" In these ways, Lenffer describes FEAT as a homegrown mission with multiple bottom lines, a pooling of resources to do things ourselves that otherwise would not be done. "It's kind of punk, really."

Lenffer is not the first to notice parallels between solar energy and punk rock. Solarpunk is a thing. It is a genre of speculative fiction that imagines the alternative futures solar energy may bring. If solarpunk began as a literary form, like its precursors cyberpunk and steampunk, it is equally true that political sensibilities of various kinds have consistently expressed themselves through the material form of solar energy—from feminist projects in the 1970s to contemporary struggles over Indigenous energy sovereignty. Yet, like all punk, the solar manifestation confronts familiar tensions between ideals and industry. Similarly, the solar variety of climate solutionism raises familiar questions about ecological principles versus economic power.[3]

This situation is best understood by thinking beyond electricity generation alone, shifting the terms from solar as a type of energy to "solarity"—"a state, condition, or quality developed in relation to the sun, or to energy derived from the sun." From this perspective, solarity makes promises. But it also tells lies.[4]

The first-order promise of solarity, as with public and corporate enthusiasm for renewable energy more generally, is that photovoltaic cells will "make the world made by oil possible after oil." In other words, renewable energy's horizons of expectation are currently being formed by desires that were cast in the era of nonrenewable energy. As a type of relationship to the sun, this form of solarity is a revitalized species of fossil capital and its petroculture. Even in China, the world leader in solar panel production and solar energy capacity, those astounding and state-led gains have been achieved alongside private corporations pursuing market share under the pressure of global capital. This means that the good news of decarbonization is being realized the old-fashioned way. Exploitation. Making solar panels is not just hard work. It can be poorly paid, dangerous, and forced labor as well as downright plunderous both inside China's borders, in provinces like Jiangsu and Xinjiang, and outside them, along supply-chain sites like Indonesia and the Democratic Republic of the Congo. Meanwhile, overall emissions keep rising. This promise of solarity, it turns out, is made to be broken.[5]

The real promise of solarity, according to its advocates, lies not in a set of institutions that continue to erase realities of exploitation and inequality. Nor is it found in a focus on carbon alone, which continues to embrace magical thinking with regard to energy transition. Rather, the promise of solarity is that, "like all mediated conditions," it "names a condition and terrain of contingency and political struggle" that influences "the sorts of subjects we might become." The hope is that these new solar subjects might be more revolutionary than reformist, that they might realize new solar societies, new solar obligations, new solar encompassments. The goal is to figure out how the star at the center of our solar system may finally be received not as a free gift of nature, as capital would have it, but as a source of value whose use and meaning are decided together and geared toward human freedom.[6]

Until then, we will continue to struggle, and seek substance, in the friction between the promises of solarity and the lies of green capital.

Lenffer is well aware of the contradictions that animate climate initiatives like FEAT. She has seen the intended snipes in online comments sections (*Oh, fantastic, greenwashed capitalism for artists!*) and she's heard the attempted gotchas about working for change from within existing economic arrangements (*Ah, yeah, you're participating in the system and just trying to get a piece of the pie for artists!*). Lenffer has considered all this and more. It doesn't keep her up at night.

On the first point, the FEAT founder wonders about the tendency among climate activists to ostracize those who, though they may not be engaged in the same kind of activity, are broadly pulling in the same direction. They say it takes all kinds to make a world. Surely it will take all kinds to remake one too.

On the second point, Lenffer says yes, of course, getting part of the pie is what FEAT was designed to do. Being involved in the financial world? That's what an investment scheme is. In addition to offering the possibility of monetary stability to musicians, those famous early victims of the gig economy, FEAT is also a response to inaction on the part of the Australian government. Lenffer condemns her country's lack of clean energy technology rollout over the decade preceding the launch of FEAT, and she makes no apologies for stepping up where Canberra has not.

If atmospheric accounting is a defining feature of climate common sense, certain philosophers suggest that a defining feature of the modern age is something called atmospheric explication. This is not explication as explanation. Rather, explication here is defined as "a historical process in which implicit assumptions about life and the environment are forced to become explicit objects of representation and management."[7]

Unlike the adage about fish not grasping the wetness of the ocean because it is all they have ever known, there is today little about humanity's carbonated atmosphere that can be taken as given or unchanging. In fact, it is increasingly recognized that environments are media of knowledge. Changing our milieux, as fossil capital is doing and as solarity hopes to, is to change the parameters of thought and action that allow us to conceptualize and politicize the climate crisis in the first place.

When it comes to the situation of our climatic modernity—a situation that capital has created and now openly exploits again—atmospheric explication is as good a description as any. To keep undoing the downward spiral, whether through solar reforms or revolutionary solarities, one goal must be to take control of the explication process. One goal must be to make it our own by upgrading the carbon intuitions and climate institutions that define our hothouse planet and constrain solutions to the crisis—as well as the definitions of climate and crisis. FEAT's story is one such atmospheric explication. James Dove's work with ClimateEQ is another. Jen Cregar's work with Terra Lumina is another. There are many more.

Thoughts like these held my attention in the Blue Mountains during the southern winter of 2022. As we stood on that sandstone plateau, overlooking the eucalypt valley where for a million years the atmosphere has naturally explicated itself in a gauzy blue gray, like a cough drop fog, Lenffer was already decommissioning FEAT. She wasn't giving up. She was moving on to other opportunities—and she was keeping the name. But Lenffer was winding down the investment side of things, partly because it had become difficult to raise money during the severe downturn in musicians touring around 2020. If fire once took her home, it was a virus that shook her fund.

The Brigalow solar farm, meanwhile, was humming along—as it will do for another twenty-some years, explicating the explicated sunlight.

11
EarthPercent

Heidi Lenffer may have considered a climate charity and gone another way. Not Brian Eno.

Eno, a self-described nonmusician, had seen the problem with the climate, and he was convinced that music could be doing more to solve it. There was a willingness among musicians to engage, Eno thought. But there was uncertainty on how they might do so. Eno had also read that only 3 percent of worldwide philanthropic contributions were being put toward climate mitigation. That seemed low.[1]

Around 2019, music industry managers Adam Callan and Hiroki Shirasuka were dealing with similar feelings. "It was obvious," they say, that it was time "to change the way we conduct our business." And they were not just thinking of themselves: "There's a lot of money in the music industry, and there's people doing essential environmental work who need financial support. We wanted to find the best way to leverage the music industry to support this vital work." Since music's revenue streams are already tapped and distributed among multiple channels—record sales, touring profits, publishing royalties, agency fees—it seemed logical to introduce a further diversion toward eco-philanthropy. All this culminated in 2021, when Callan, Shirasuka, Eno, and others established EarthPercent, "a simple way for the music industry to support the most impactful climate causes."[2]

EarthPercent is a charitable organization, which is an institutional form that has overlaps with nongovernmental and nonprofit bodies. The charity asks individuals and companies from across the music industry to commit a small portion of their income stream—the "earth percent"—to the cause. It is even possible to credit Earth as a songwriting partner and to donate the royalties. (Many have claimed nature as an inspiration. Few have named nature as a coauthor.) Once a sufficient amount of money has been raised, EarthPercent publishes a call to use the funds, vets the applications, and establishes its grant partners.

Those grant partners fall into four categories. EarthPercent funds projects initiated by groups directly experiencing the effects of climate change. It funds "solutions that science has told us we need to invest in." It supports scrutiny of the "hidden wiring" (laws and policies) that can hinder or foster system-level transformation. And it supports "changing cultural norms and narratives." As of 2024, the charity had named its first sets of grant partners and disbursed over $1 million. By the end of the decade, it aims to have raised and redistributed $100 million.[3]

The intended recipients are the organizations addressing climate crisis with the most ferocity and urgency. Although EarthPercent's self-presentation emphasizes the climate, and thus carbon, the charity is thinking beyond emissions—and beyond music. For example, some of their first round of funding went to music-climate mainstays like Julie's Bicycle and Music Declares Emergency, but funding also went to the Global Greengrants Fund, which supports Indigenous communities to organize against fossil fuel exploration and extraction. Money went to Women4Oceans, which works against overfishing, pollution, and deep-sea mining. And support was offered to Cool Earth, which works with rainforest communities to defend against deforestation.

Part of EarthPercent's funding model is reminiscent of a tax or even a tithe. The model is not new. Nor is it exclusive to the organization. In fact, Heidi Lenffer has piloted a broadly similar initiative called FEAT.Live: The Solar Slice. The Solar Slice is a negotiable 1.5 percent ticketing surcharge "that will fund crucial carbon reduction measures for the live music and entertainment sector." It is meant to "bake sustainability into the DNA of all live events and to provide the money that's necessary to prioritise serious climate action."[4]

Both EarthPercent and the Solar Slice are in relatively early phases. Still, the seriousness of their commitments and the scale of their ambitions are clear. They are putting climate crisis at the top of music's political agenda. "We're trying to make it *the* cause of the music business," Eno says. "Let's have a revolution."[5]

These are inspired initiatives. But there are hiccups. And this is not just because essentially taxing musicians, most of whom already have a hard time making a living, or passing costs along to fans, most of whom already have a hard time getting by, are strategies that quickly fall into conservative traps. There is a tendency for right-wing ecoliberalism to smear left-wing ecoliberalism as a plot to make everything more expensive—which, often enough, is a fair criticism. The deeper issue is that the revolution will not be philanthropized.

Along with various other nonprofit, nongovernmental, volunteer, and civic organizations, philanthropy constitutes an economic force known as the third sector. The third sector falls between the public and private and is geared toward the advocacy of social causes and the provision of social services covered neither by governments nor businesses. It all sounds pretty innocent and good natured, and generally it is.

Yet the third sector has a complicated history with regard to social causes. Institutions like the Rockefeller, Ford, and Mellon Foundations, for example, have not only steered certain progressive social demands toward more conservative ones. They have also contributed to the violent repression of various radical social tendencies. The Black Power, Red Power, and Third World movements of the 1960s and 1970s are some of the best-known examples here. But why should this be?[6]

One reason is that it is in the interest of such institutions to ensure their own longevity more than the liberation of their causes. This is by design. In their modern form, nonprofits were established by the wealthy as tax avoidance schemes beautified by the kind face of charity. By virtue of their relationship to capital, then, philanthropic bodies appear progressive and certainly do some good. But they treat their underlying causes as chronic rather than curable conditions. Like greenwashing, this is no one's fault. It is a system-level feature. Charities are caught between their immediate and fundamental interests.

As the third sector has taken further shape since the 1970s, filling the increasingly large gap between the social welfare provided by the public sector and the social welfare pilfered by the private sector, philanthropic institutions have become more necessary and self-perpetuating. They have "taken the form of social control mechanisms to manage the social unrest and disorder that results from the dismantling of the social safety net." The redistributive model sprinkles financial resources on the surface of relief, without disturbing the underlying reality of economic power. All these factors lead critics to describe the third sector as the nonprofit industrial complex.[7]

Climate philanthropy appears here as yet another ecoliberal variation on a wider neoliberal theme. In addition to devolving political accountability onto the personal and the corporate (the responsibilization of the climate), ecoliberalism devolves social welfare onto the charitable (the responsibilization of the crisis). It is yet another site that accommodates that devilish dance between ecological principles and economic priorities, that delicate ballet of resistance and resignation under the spotlight of economic power, that sympathetic symbiosis among consumer activists and conscious capitalists. Climate philanthropy may be just the next industry, or para-industry, that exists for its own sake—and to ease the transition to green capital—more than anything else. Another answer poised to turn back into the question it promised to address.

That is the risk. But there is potential here, especially for those who insist that the problem is not climate or crisis but capital and class. What if charities, nonprofits, and NGOs did even more to fund the salaries of people who organize protests, workplace solidarity efforts, and so on? What if they channeled their resources toward an international strike fund?

Such causes may seem remote from climate action. They require explanation, because they will not directly reduce carbon emissions, protect a forest, or save a species. The connection to climate is indirect, but the results are equally significant and potentially more so.

As the face of EarthPercent, Brian Eno knows that funding indirect causes can have substantial leverage in the struggle for the planet. Eno believes that finding ways to solve the climate crisis will be tied to the alleviation of other social inequalities and injustices. "The problem is big," he says. "But the reward is bigger."[8]

Sarah Ditty agrees. Ditty chairs the expert panels who vet applications for EarthPercent funding. Her day job is with a philanthropic foundation that is involved in funding trade unions and labor organizations. Some of those organizations are "building greater solidarity between workers in Global South countries through workplace organizing."[9]

Funding worker solidarity efforts and strikes is to acknowledge and support class politics, which is the most powerful way of confronting capital—the source of the climate crisis. It is a step toward not only redistributing money but also reconfiguring economic power. It is a hopeful indication of the difference that organizations like EarthPercent can make and of what the music industry more generally is capable of.

Hamburg has some of the finest dive bars. In September 2022, I was looking for a place to write some notes, somewhere neither too loud nor too crowded. It was easier said than done. As happens every fall, this part of the city has been transformed into one big music industry carnival. There are at least 40,000 attendees and more than 400 concerts over the four-day Reeperbahn Festival. Everything in the vicinity is crammed with hyper-stylish musicians, big-business talk, and lanyards. Lots of lanyards.

The festival has been held in Hamburg's St. Pauli neighborhood since 2006. Although it is certainly a music event, Reeperbahn today is also reminiscent of South by Southwest. It is not quite a festival, not quite a conference, but rather some kind of looser entertainment industry networking gala that features plenty of concerts by musicians

Reeperbahn Festival Sustainability Hub in Hamburg, 2022.

seeking record deals and the currency of exposure as well as a program of film, visual art, virtual reality, literature, awards, panels, and more.

Reeperbahn also has a dedicated Sustainability Hub. It may be just a small room off the lobby in the main conference hotel, but it hosts a variety of climate-oriented events over the four days of the festival. This includes an EarthPercent presentation, which I was invited to join on a panel with a title that varies the charity's tagline: "Unleashing Music in Service of the Planet."

Our chair is Golchehr Ghavami, an agent and manager in electronic music as well as a creative ambassador for EarthPercent. We are joined by Sarah Williams, a music and entertainment lawyer as well as a member of the EarthPercent publishing committee, and Wayne Snow, a musician who appears on EarthPercent's inaugural Earth Day compilation album. (The second edition features a limited release on Evolution Music's bioplastic.)

Among the various conversations across the panel, Snow spoke movingly about being a parent in uncertain times, about touching people's hearts through music, and about how people need to get right mentally, almost spiritually, before they can address imposing issues like climate crisis. Ghavami called for carbon-monitoring music technology—ways of knowing and assessing how much we emit from streaming, buying albums, attending concerts—and she lamented that such monitoring systems were not already standardized and set in place. Williams talked about the legalities and benefits of crediting Earth as a songwriter, while an audience member made a connection to tithing—which resonated in Germany, where the church tax is 8 or 9 percent of an income.

It all went well enough, in my memory. Still, it was the pre-panel meeting, and the post-panel debriefing, that I remember most.

At the online pre-panel gathering, we rehearsed some of our talking points. My job was to describe the environmental consequences of the record industry and to talk about the good work of recomposition. As the conversation developed, the question of underlying problems arose. Capital, I suggested, was the culprit. This was rightly brushed aside. It was pointlessly obvious. Everybody knows it's capitalism. Focus instead on what is realistic to address and accomplish. Exemplify the pluses, not the minuses. Concentrate on solutions, not the problem.

Nevertheless, capital did come up during the panel itself. So the topic wasn't forbidden, but it did not energize our public conversation. It short-circuited it, more like. It was a basic misunderstanding—at a business event—to say there was a problem with the goose when everyone was there for the golden eggs.

Afterward, though, during our post-panel debriefing, the conversation was differently charged, precisely because of capital. The economic arrangement may have been our pointlessly obvious problem, but offstage it sure was good to talk about it. Considering capital opened all kinds of big questions. Questions about a person's relationship to the work they do, about what makes life worth living, about financial problems and health problems and family problems, and about how much easier it would be, especially after the hardships of recent years, to just give up on all this climate stuff and make some real money.

The tension between economic priorities and ecological principles is rarely so raw and exposed. Rarely does it appear so frayed. Once noticed, the same basic tension, the same fray, seems to be everywhere these days, in ways both specifically individual and broadly social, between the personal and the political, in that place where the risk of resistance rivals the realism of resignation, where the possibility of hope holds a candle to the convincingness of despair.

It was cathartic.

The Reeperbahn Festival hotel spills onto Hamburg's Spielbudenplatz, an area that has been home to entertainers for over 200 years. These days it hosts food trucks, beer gardens, night markets, dog fairs, biker rallies, and, of course, concerts.

After the EarthPercent panel, and the debrief, Reeperbahn's Korea Spotlight was winding down on the platz. The spotlight featured a musician called 250. He played a DJ set from his award-winning debut album, which revitalizes an old Korean popular genre, *ppongjjak*, using contemporary production.

Ppongjjak's sentimental sad-glad melodies and danceable one-two rhythms may be "like oxygen," as 250 puts it, in that they are everywhere in South Korea's popular music atmosphere. But 250's album is at the same time "a lonely work," according to one journalist, because it touches on turbulent Korean histories. "From the moment the synthesizer's delicately placed first minor chord hits your ears, there's no finding your way to a cheerful dance."[10]

As the set ended, I continued along the Reeperbahn, reflecting on other atmospheres, other turbulence, and other dances. Hamburg's dive bars may be inviting and memorable, but I settled somewhere nonsmoking and forgettable, next to the Steak Fish House, around the corner from the places men distribute fliers for €59 favors plus a free shot with your first drink, and typed out some notes.

12
Low-Carbonism Music

Carbon is the lingua franca of the climate crisis. It is also the common currency of that crisis and its standard unit of measurement. Meanwhile, *Homo carbonicus* encapsulates the common sense of it all—offering cultural equipment for living, giving shape to collective consciousness.

In media research since the mid-twentieth century, the historical composition of consciousness and culture has been written in three movements. For a long time, the story goes, humanity was a creature of oral consciousness that lived by its ears, in sound. Then, with the printing press, humanity became a creature of literate consciousness that lived by its eyes, on the page. Twentieth-century mass media, like radio and television, called for a new hybrid orientation of this creature to its world. This gave rise to a form of secondary orality, along with the electronic consciousness that went with it.

Not all midcentury researchers of media and culture adhered to this schema. Yet most of them recognized they were living through a significant shift in the conditions of consciousness and culture, and that the uses of this new media literacy were neither given nor settled. Developments in mass communication made for "both a moment of danger and one of opportunity."[1]

These developments were dangerous because they represented a time when "the new dynamic forms of mass culture" signaled "the incorporation of the masses more fully into the market" in addition to the erosion of established cultural practices. The developments were opportunities insofar as this was also a moment when new media and their mass cultural forms were evidently opening up novel, against-the-grain, subcultural rituals of resistance (and resignation).

It would be too much to suggest that *Homo carbonicus* represents some grand new epoch in the history of humanity. But carbon is undoubtedly on everyone's mind, and so the figure does seem to say something about the interpersonal equipment of consciousness and culture today. If orality is in sounds and ears, while literacy is in texts and eyes, and if electronically mass mediated popular culture mixes the two, in a form of literate orality, then perhaps carbon consciousness consists in the media of breathing—in air and lungs. Or maybe it is in the media of thermoception—in radiation and skin. Either way, like the mass media moment of the mid-nineteenth century, our own elemental media moment is full of danger and opportunity.[2]

Carbon literacy, carbon metrology, carbon credentials, carbon citizenship—there is no denying that we need all of this, and more of it, to some extent. In music, it is clear that important institutional work is being carried out by numerous individuals and organizations. The promise, the moment of opportunity, is plain to see.

The dangers of such a schema—human history as a suite of movements from oral to literate to electronic to carbonic consciousness—are equally apparent. This tradition of research has been unwittingly based on "antiquated notions of sensation and cultural difference," says researcher Jonathan Sterne, which "posit the ascendancy of the White, Christian West as the meaning of history." In other words, the beliefs about human progress and human values that guide such work have been based on an ethnocentric teleology. A carbonic vision of our species may not echo the same kind of damaging evolutionary politics about humanity's time on earth, one where there are effectively two versions of the cultural present that are separated by a developmental gulf. But the addition of carbonic consciousness to this schema is doing similar work. Except, instead of resting on the construction and subjugation of a temporal other, it rests on the construction and subjugation of a classed other.[3]

What is carbon? Researchers like Anne Pasek have spent a lot of time asking this question, "thinking about how we think about carbon." One of Pasek's conclusions is that carbon is not a thing. It is a relationship. This means "the element's essence is best parsed by studying its bonds." Something similar can be said for its political valences.[4]

Yet in the same way that the emphasis on *climate* in climate crisis serves to sidestep broader ecological and social issues, the politics of carbon that defines climate common sense—carbonism—also works to conceal its real bonds. While decarbonization is a pressing matter, there is a centrifugal force whereby capital disburdens itself of the true costs of low-carbon energy, imposing those messy externalities of emission and extraction onto ever poorer people in ever poorer places.[5]

It is important to ask what carbon is. Some historians of science believe the key line of inquiry is actually about "who" carbon is. In addition to those questions, we should be asking another. What does carbonism want? The question contains an important ambiguity. To ask what carbonism wants is not to ascribe vitality or purpose to carbon, or to wonder about the physical tendencies in the life of an element. It is to question what carbonism lacks. It is to ask what carbonism is missing as a political vision. Or, as the expression goes, it is consider how, having been weighed in the balances, carbonism has been found wanting.[6]

One goal of carbon literacy should be to restore this element, and the elemental thinking surrounding it, to their social and historical circumstances. Otherwise, carbon neutrality will continue to disguise many of its true effects. It will remain a corporate production, affording tricks of accounting that allow *Homo carbonicus* to negate and subjugate its others. It will be a form of shorthand that annuls the complexity of the world in favor of another, more restricted form of complexity: number crunching.

Pasek has thought about this malignant numeracy too. "Numbers," she says, "are necessary to frame the problem of carbon as a planetary commons." Atmospheric accounting is required, not just to understand the climate but also to underline that its crisis does not respect national borders (even as it unevenly effects various people and places). Still, governing the planetary commons, and confronting the sources of climate crisis more generally, has to involve citizens and institutions that are capable of something more than the disciplined abstractions of footprints and bookkeeping. "There is much to gain," says Pasek, "in breathing equity concerns back into the ledger's figures and coming to articulate forms of measurement that acknowledge that the inputs and outputs of the carbon cycle are never as narrow as the media of its accounting might suggest."[7]

Focusing primarily on how to count carbon may even invite a shift from carbon footprinting to carbon fingerprinting, a tool of self-surveillance and state regulation to be maintained by a carbon police force. It isn't as far-fetched or dramatic as it may seem. The Glastonbury Festival, for example, albeit tongue in cheek, has sometimes featured Green Police patrolling its grounds. The goal is to make sure campers "take responsibility for their waste."

It is not just that such frameworks present incomplete, divisive, or repressive pictures of the world. Carbonism also universalizes problems and solutions that are specific to the economic core at the expense of its peripheries both abroad and at home. This is the convenience of carbon. It sustains the war with otherness that animates the capitalist climate crisis. It upholds the social architecture dictating who may be treated like a subject (us, ourselves) and that which, to continue the upshift from fossil capital to green capital, must be treated like an object (it, them).

Those of us concerned with the future of music should be concerned with the kinds of institutions that regulate and reproduce it. Carbonism is quickly becoming one such institution. Its effects on what music is, and how music comes to be, are still emerging. But a few things are clear. Our goal should not necessarily be to construct a more complex, more complete, or more caring *Homo carbonicus*. It should be to understand the simplicity, impoverishment, and violence of that figure as a historical abstraction and institutionalized form of objectification. The goal is to resist our complicity in that figure's social domination and reproduction, to find what lies beyond the convenience of carbon.[8]

As much as we need low-carbon music, we need low-carbonism music too.

If consciousness and culture have been so clearly linked in social and political thought since the twentieth century, and if it is clearly necessary to reshape both, not to solve the climate crisis but to confront its causes, then it is sensible to ask where change begins. To answer that question, a bit of folk wisdom might help. This wisdom says it is easier for people to act themselves into new thinking than to think themselves into new acting. If that is true, then our best bet for change might not begin with consciousness. It may start with culture.

Here we open a set of questions that, until this point, have been largely muted. Reflecting on the projects detailed in this section, some readers might object that, in the big picture, music is small potatoes. Its contribution to the climate crisis is trivial compared to other industries. I have never thought that the relatively small size of music's climate contribution was a good argument for ignoring the fact of its contribution. It is meaningful to tend our little garden. But we should also be focusing on the real polluters.

The music world agrees. Those involved in the recomposition spend at least as much effort on music's adaptation to climate crisis, in terms of technical and institutional solutions, as they do on music's role in relation to more general climate mitigation strategies. They want to know how to make music more sustainable, how to solve music's own relationship to climate crisis. And they want to know how music can help solve the wider climate crisis.

As the recomposition pivots, so do the questions we have to ask about it. What about the music on the ecological records that Billy Bones and others are pressing, not just the stuff those technologies are made of? What about the songs by those artists who have donated their profits to build solar farms or who have given their time to become carbon literate, not just the institutions of solarity or literacy? What about content, not just form? Messages, not just media? What about music as music? What about cultural solutions to the climate crisis?

Part III

Cultural Solutions

13
The Missing Link

"Culture," says Alison Tickell, shortly after taking the stage, "needs to be part of every single conversation and every single solution to the climate crisis." She has my full attention.

Tickell is introducing an event hosted by Julie's Bicycle, the UK nonprofit she founded in 2007. In its early days, Julie's Bicycle described its "simple primary purpose" in terms of supporting "climate responsibility and carbon reduction" in cultural industries and arts councils. One of its initial undertakings was to produce a report on the carbon footprint of the UK music industry. The report was a first of its kind. Its influence has been widespread.

Today Julie's Bicycle has a broader remit. In addition to focusing on emissions reduction in the cultural sector, which they still do through projects like the IMPALA Carbon Calculator, the organization has expanded its mission. Julie's Bicycle now devotes at least as much energy to "mobilizing the arts and culture to take action on the climate and ecological crisis." The event Tickell is on stage to launch is called We Make Tomorrow, and it is clearly part of her organization's expanded mission.[1]

We Make Tomorrow at Centenary Square in Birmingham, 2022.

We Make Tomorrow took place at Birmingham's Centenary Square, on a sunny day in October 2022. According to promotional materials, attendees could anticipate "an action-focused one-day summit with performance, conversations, and workshops to celebrate and learn from creative climate leadership." The event website, meanwhile, promised to build on an earlier gathering by the same name, which took place in 2020, where "the cultural community mobilized on climate action, and connections between environmental, social, and cultural justices were exposed." Now, at the event's second iteration, the goal was to "explore how working with shared purpose can generate social, economic, and creative value that helps us all to imagine, and craft, a better tomorrow."

All that to say, I was not quite sure what to expect when I arrived in Britain's second-largest city on that pleasant autumn day. But I knew why I was there.

A year earlier, Julie's Bicycle had published a report called *Culture: The Missing Link to Climate Action*. The report itself is directed toward policymakers. Partly, Julie's Bicycle was recommending that "the arts, culture, and creative industries" should be more stringently subjected to "national statutory requirements to meet and contribute to climate targets." In other words, they were asking for stricter carbon monitoring and reporting requirements in the cultural industries.[2]

Julie's Bicycle has led the way here. This is especially visible in their collaborative work with Arts Council England. Since around 2012, according to the organization's CEO, Darren Henley, the Arts Council has been "embedding environmental reporting and planning into funding agreements." Henley stresses that the Arts Council is "the first cultural funding body, anywhere in the world, to make a commitment to the environment a requirement for our regularly funded organizations." Many others are following suit.[3]

In addition to looking inward, to regulating and evaluating cultural activity according to climate criteria, *The Missing Link* report is also directed outward, to the idea that "cultural policy can strengthen the creative climate movement, and thereby mobilize action at scale." Which is to say that, on top of climatizing culture, Julie's Bicycle is also advocating the culturalization of the climate.[4]

This is the idea that most interested me. It featured prominently in Tickell's opening remarks at We Make Tomorrow. She was speaking partly in anticipation of her organization's attendance at the 2022 United Nations Climate Change Conference, or COP27, which was being held the following month in Egypt. This was not the first time Julie's Bicycle would be present at a weighty UN gathering. They were there at COP26, for example, when our Glaswegian protester reminded everyone of the true need for such conferences. (It's capitalism, ya eejits.) But COP27 would be special.

"For the first time," Tickell announced in Birmingham, with evident and justified pride in the achievement, "culture, the arts, heritage, and creative community will have formal presence in the room" at such a high-profile event. Thanks to her work, culture has found its way to the most official channels and the very top levels of international climate negotiations.

Culture's ongoing presence in all this is assured. It was cemented at COP28 in the United Arab Emirates, where dozens of national ministries, intergovernmental bodies, and cultural organizations from around the world unanimously adopted the Emirates Declaration on Cultural-Based Climate Action. The declaration insists that any serious effort to meet the Paris Agreement "must include a focus on cultural heritage, arts, and creative industries and the sociocultural enabling conditions for transformative climate action." There were plans to strengthen this commitment to culture at both COP29, in Azerbaijan, and COP30, in Brazil.[5]

The Missing Link report is part of the fabric of these developments. Its main message and its game plan are woven through culture's newfound prominence in climate common sense. "Cultural policy," the document declares, "can embrace narratives that link art and culture with the overwhelming imperative of climate transformation." Then the wager: "At scale, this might just be the missing link."[6]

The missing link. It sounds impressive, an environmental equivalent to discoveries in the fossil record promising answers to the grand mysteries of who we are, where we come from, and what it all means. If humanity is the only species with an ego big enough to see its own existence as a problem to be solved, maybe climate crisis is an existential threat big enough to fast-track the process—for better (edification) or worse (extinction).[7]

Of course, the missing link report is not given to defeatism. It wants to help, and it points the way. Like many documents of its kind, though, *Culture: The Missing Link* is written with a bureaucratic accent. It can be challenging to find the foundation of meaning in the quicksand of jargon. What exactly would it denote, for example, to embrace narratives that link culture with climate transformation? What is this previously missing link actually linking? What does the link look like? How is it supposed to work?

The closest thing to an answer comes late in the report, where the underlying mechanism is suggested. Synthesizing survey results from thirty-one national cultural ministries and arts councils, from Argentina to Zambia, who were called on to answer questions about the "the role and function of the arts and culture in relation to climate change," the Julie's Bicycle report suggests the main job of culture is to "raise awareness."[8]

Many synonyms appear in the document. The arts and culture can "place the climate issue in public discussion." They can "inspire" us. They can help us "change perspective" and "envision a positive future." They offer a "unique leadership voice." And so on. The point is clear. Culture is understood as a crucial but missing link between climate awareness and climate action.

This is what Julie's Bicycle describes as "the power of culture." It is a power that can be "harnessed" in the interest of "mainstream climate action." They are not alone in this language. Broadly similar organizations express the same ideas in the same terms. Creative Carbon Scotland, for example, believes that culture has an "influencing power," which they are working to "harness." Likewise, Green Music Australia is "harnessing the cultural power of our influential music scene to create a greener, safer future." In the United States, an organization called Reverb also harnesses the power of music to move us "to feel, to care, to take action."[9]

The basic idea here is obviously widespread. Julie's Bicycle has played a pioneering role in developing and publicizing it. They represent the leading edge of how culture is being rethought, reevaluated, and reorganized in relation to the climate crisis. It all sounds so intuitively and unassailably positive.

But what is the underlying state of affairs that makes this idea possible? What allows it to seem so obvious, self-assured, and true?

Some of the answers can be gleaned from *The Missing Link* report. Some were on display at We Make Tomorrow. Other answers have to be found in history.

The ideas about culture underpinning Tickell's opening remarks, the work of Julie's Bicycle, the Emirates Declaration—these ideas come from somewhere. Many of them can be traced to a place just down the road from We Make Tomorrow, where a blue heritage plaque gleams on gray brutalist concrete. The plaque commemorates the Centre for Contemporary

Cultural Studies, which was founded a short drive from Centenary Square at the University of Birmingham in 1964 and is an essential site for understanding a set of epochal shifts in the foundations of the world going back to the middle of the twentieth century.

Those shifts—including radical changes in political economy, progressive philosophy, and the sociology of the academy—cannot be reduced to the ambush of neoliberalism, the rise of the New Left, or the explosion of cultural studies. Although those are important plot points, there is no single storyline here, no individual sequence of cause and effect. Still, one thing is sure. From the middle of the twentieth century, something called culture moved to the foreground of social life and to the center of the public sphere in ways that it hadn't been before.

This was not your grandparents' culture. It wasn't culture as the finer things in life, like table etiquette and great works of art. Nor was this the culture of the anthropologists. It wasn't just the ordinary things in life, whole ways of existence that create meaningful orientations toward the world.

The big idea here, both theoretically and politically, was that cultural practices have consequences. Such practices were no longer seen as by-products of an economic arrangement or its social architecture. The economy and society were themselves increasingly understood to have been produced and maintained less by material interests than by values, meanings, interpretations, and imagination—in a word, culture. To use a favorite sign of the times, culture was understood to "mediate" the relationship between a person's interests and a person's politics, and so to mediate the reproduction of public order more generally.

All this rose to prominence partly because it helped explain an anomaly. It helped explain why, although workers had been widely regarded as the primary agents of progressive historical change until around the Second World War, the class had not organized itself and achieved a revolution in the way some leading social theories had predicted. History, it seemed, had gone haywire.

If shop floors and mine shafts were apparently not the places where classes in themselves would become classes for themselves, maybe the secrets of class formation would be found in leisure rather than labor—in everyday cultural activities like reading magazines, listening to the radio, and watching television. The idea made sense not only in a mid-century moment when new forms of mass media were taking hold. It also shed light on a time when workers, whose lives had for a brief postwar period been lifted out of poverty, were by the 1970s being gradually but brutally decomposed as a class. This had a crushing effect on organized labor, helped make room for a proliferation of new social movements, and coincided with the rapid expansion of postsecondary education in many countries—meaning that leftward thinking found a new home in the academy.

Previous social theories, these intellectuals began to argue, had been reductive and economistic. Earlier thinkers had been overly committed to the labor metaphysic, to workers as the necessary lever of history. But the world had changed. What was needed were new theories that responded to those changes—theories that understood all the ways that power, politics, and people took shape outside the sphere of production. What was needed, it seemed, was not economism but culturalism.

The developments and debates of these long, global 1960s were encapsulated in the Birmingham Centre for Contemporary Cultural Studies. All in all, they are referred to as a cultural turn in twentieth-century thought and politics. The cultural turn is said to represent an inversion of previous thinking. Here the explanation for social stability and social change is less the domain of historical materialism (where reality precedes and shapes thought) and more the domain of philosophical idealism (where thought precedes and shapes reality). The possibilities for acting in the world are flipped. They become less about organized labor coalitions than swarming multitudes of protest movements. Less about groundwork than imagination. Less about political economy than cultural politics.[10]

It is this history—this structural inversion, this so-called turn—that Julie's Bicycle and many others are invoking today when they suggest that culture is the missing link between climate awareness and climate action.

I never did make the pilgrimage to the plaque. This was despite the fact that the University of Birmingham's Muirhead Tower, where the heritage marker hangs and where the Centre for Contemporary Cultural Studies was once housed, is itself an attraction. A beautiful brutalist monument. Still, other landmarks for other hometown heroes, like Black Sabbath Bridge, seemed more fun.

Walking around town, though, I couldn't help thinking about the center and the tower. I thought about how their architects, Stuart Hall and Philip Dowson, both passed away in 2014. And I thought about another coincidence—about how academic culturalism and architectural brutalism have each been seen as emblematic of both modernity and postmodernity, as signs of all that is right and wrong with the world. Yet the version of the story presented here so far, which was for a long time the only version I knew, emphasizes only the goodness and necessity of the cultural turn. You wouldn't know from events like We Make Tomorrow, reports like *Culture: The Missing Link*, or the plaque on the wall that there has always been a substantial left-wing critique of culturalism.

The left's critique of culturalism begins in a series of ironies. In the same moment cultural studies is celebrating the subversive potential of everyday life and disorganized acts of consumption, capital is remorselessly organizing and expanding the sphere of production in its own image. In the same moment cultural thought begins retreating from class politics, capital's resolute commitment to class war is growing stronger. In the same moment cultural politics is finding inspiration in slogans about the power of knowledge and the imagination, capital is relentlessly demonstrating that power flows first from the economy. And in the same moment cultural theory is becoming suspicious of structures, totalities, and universals, capital's systemic unity is becoming more total and brazen.

Some commentators saw and worried about all this from the get-go, including culturalism's sharpest minds. By the new millennium, the limitations of the cultural turn were so obvious that some exasperated onlookers were reduced to cracking jokes about the descent of cultural studies into dissertations on "the structural analysis of beermats" and "the phenomenology of lying in bed all day." For others, culturalism was no laughing matter.[11]

Adolph Reed Jr. is one of the left's most fearless advocates. He is also one of its fiercest critics. In a long career focused on the politics of race and class in the United States, Reed has always been dead serious about culturalism. He disparages cultural politics as playacting. He attacks the cultural turn not only as a distraction from more serious matters but also as a superficial and repressive project of the upwardly mobile. "Despite the dogged commitment of several generations of cultural studies graduate students," Reed says, "who want to valorize watching television and immersion in hip-hop as politically 'resistive,' it should be time to admit that that earnest disposition is an ersatz politics." For Reed, a politics of culture is worse than no politics at all.[12]

Talk about this version of events with people invested in the cultural turn, and you are likely to get a pfft of dismissal. The criticisms are reductive, they'll say, because cultural theory is not about replacing economism with culturalism—it is about superseding the binary altogether. The criticisms are unfairly targeted, because cultural studies have never been just one thing. Or the criticisms are wide of the mark, because cultural theorists have contributed lots of compelling ideas, cultural politics has achieved many significant wins, and fields of inquiry should be judged on the best of what they yield. The criticisms may eventually induce a yawn of indifference, because culturalists have heard it all before and simply don't recognize themselves in this story.

Most of these are reasonable points of contention. Certainly, in this light, the comments of someone like Reed can be divisive. Yet they cannot be dismissed. How can culturalism be everywhere in general but nowhere in particular? Should fields really be judged by their outliers and not their standard distribution? If the debate is so dull and settled, why is it heating up again? And if the critical version of the cultural turn is so far from reality, then why does it make such complete sense of my own experience?

I am a product of the cultural turn of the 1960s. Most of my professors either had a hand on the wheel, were taught by those who did, or both. I even have a degree in missing link-ology, or the study of "cultural mediations."

I am also a child of the 1980s. This means that I am one of those millennials for whom what observers call late capitalism, neoliberalism, postindustrial society, the post-democratic age, the fifty-year counterrevolution was just how things were for me growing up. It was the way of the world. It was everything.[13]

By the time I started college and university in the early 2000s, culturalism was everywhere and curriculums were influenced accordingly. The main lessons we were taught, as I recall them, were about overcoming snobbery, understanding identity, and celebrating resistance—whether resistance meant consuming culture against the grain or producing artifacts that opened their audiences to some kind of disruptive effect.

We learned that distinctions between high culture and low culture, while real enough in their effects, were aesthetically arbitrary and untenable to argue. As such, we were emboldened to study the importance of the everyday, because that's where most people spend most of their time. We were empowered to defend the importance of popular culture, because that's the culture that most people participate in most of the time. It is the realm where most people craft their senses of self and senses of community. It had to be taken seriously.

All this took place in fields of power and politics. Some selves and some communities were formed and expressed more easily than others. It was important to understand those processes of differentiation and discrimination—as well as to spotlight their correlates in stylized nooks of dissent and pockets of freedom. In the end, it was clear that high cultural forms like classical music worked this way too.

None of this is bad or mistaken. After Birmingham, though, as I was coming to terms with the critique of culturalism, I was compelled to sit down and reread some founding documents. What I discovered there surprised me. My own encounters with the cultural turn were distant from its initial political mission, including in music research.

That political mission certainly existed in cultural studies' early days. An essay from 1975, for example, offered the usual defense of working-class and media-saturated teens as active meaning makers—"the fact that young people are heavily involved in commercial institutions does not mean that their response is simply a determined one"—while also concluding that "the final question is how to build on that culture, how to organize it, transform resistance into rebellion." Other music research from the same period began from a different emphasis—the relationship between music, protest, and propaganda—but ended up in the same place: the question of how all this was related to the formation of class consciousness.[14]

Where was this focus in my own training? Had I simply missed it? (It is, after all, the luxury of the middle class not to have to notice class.) Had my education let me down? Could it really be that cultural politics, which seemed so vibrant and important—and which is clearly at the root of today's musical recomposition as well as wider political forms of climate common sense—amounted to little more than theater?

By the 2020s, an extensive reevaluation of the cultural turn was underway. This includes work by scholars who have lived through the period since the 1960s. Some of these scholars, at the sundown of their careers, have published retrospectives trying to make sense of it all (and in some cases where it all went wrong). The reevaluation also includes a number of youthful leftist publications and podcasts, which are taking stock after the recent explosion of international uprisings stretching from roughly the early 2010s (the financial meltdown and the Tunisian revolution) to roughly 2020 (the defeat of major leftist forms of populism and the end of lockdowns)—uprisings that were broadly based on ideas and tactics inherited from the cultural turn.[15]

From the long, global 1960s to the long, global 2010s. These episodes bookend the roughly half century of shifting consensus paradigms in political economy, progressive philosophy, and the sociology of the academy that gets referred to as the cultural turn.

The cultural turn achieved some remarkable things. It also has some noticeable limitations. There is a tangible sense that we are in a moment of reckoning with both. This moment is not about nostalgia for what went right in the 1960s, nor is it about dismay for what went wrong in the 2010s. The point is to learn lessons that may shape new politics. It is to ask: Knowing where we have been, what next?

14
We Make Tomorrow

Ask anyone how music can help solve the climate crisis, and the answer is likely to vary a theme straight out of the cultural turn. Musicians have platforms. Artists have voices. Music communicates. It touches our hearts and minds. It binds us. It moves us. Music has a power that, properly harnessed, can provide the missing link between awareness and action. Through music as culture, we make tomorrow.

Along with Julie's Bicycle and the culturalists, musicians have taken a keen interest in these capacities and possibilities. In such work, both artistic and academic, music is understood to form our identities and imaginations. It can bring about novel ideas and stitch together various ways of thinking about the world. It has the potential to make us see things differently, tell new stories, and draft blueprints for how things might be otherwise. Music may sometimes hold a mirror up to nature, as in an Edvard Grieg symphony or a Gabriella Smith concerto. But in our present condition, representing the natural world can seem less pressing than "asking what music might set in motion in terms of stimulating ecological understanding and making room for social and affective spaces of sustainable action."[1]

It's not just music. Similar ideas are expressed in wider artistic engagements with environmental themes. You can also find these ideas embraced in the branches of the humanities that have recently taken up environmental issues. The underlying conceit of this work, both artistically and academically, was recently summarized in a prominent journal of the environmental humanities. "Environmental artworks," say the authors, in language reminiscent of *The Missing Link* report, "often participate in making the distant effects of climate change sensible to a broader public, provoking open-ended affective responses."[2]

In other words, environmentally themed artworks can spark climate awareness and lead to climate action. The formative logic here is based on a set of convictions about the potential of artistic creation and human imagination. Imagination, in particular, is defined as "a way of seeing, sensing, thinking, and dreaming the formation of knowledge, which creates the conditions for material interventions in, and political sensibilities of, the world." The "most radical potential of the arts and humanities," according to this thinking, "is the ability to constitute alternative possibilities to neoliberal approaches to environments, which actively attempt to disable the imagination of alternatives." The same basic idea was in the air at We Make Tomorrow in Birmingham. As one of the hosts noted, "the climate crisis is also a crisis of the imagination, because we live in a structure that makes it impossible to imagine otherwise."[3]

Similar sentiments are widespread. It is common these days to hear that it is easier to imagine the end of the world than the end of capital. The condition is often summarized as a pervasive form of capitalist realism. There is truth to the observation. However, what musicians, artists, and humanists suggest is that we can and do imagine alternatives to capital, with ease, all the time—especially in the realm of culture and its artifacts.

Acts of imagination will always be beautiful, valuable, needed. Voices of artists will always tell truths. Works of art will always stir us. But the politics of imagination underlying such acts are tangled in a specific history of the arts and humanities.

In the early nineteenth century, one widespread way of thinking about the arts was in terms of social control. The arts were understood as positive forms of propaganda that could help shape progressive citizens. Positive propaganda became so influential as a doctrine of the arts that it has even "been credited with stimulating the growth of an entirely opposite aesthetic trend, that of art for art's sake."[4]

This romantic commitment to the independence and self-motivated nature of the arts only grew stronger through the twentieth century, especially during the Cold War, when it became bound up in the politics and operations of freedom in the First World. Rather than Second World social control and propaganda, First World arts and humanities called for jarring and disruptive works that cleared spaces for freedom of expression. They valued complexity, open-endedness, and anti-instrumentalism.[5]

The value placed on imaginative acts that has come down from the cultural turn, and that has been magnified in the distressing world of climate crisis, is not quite anti-instrumentalist. Artists and humanists believe that we make tomorrow through culture. They believe that imagination is an instrument that can set us in motion. However, there is often a notable and principled lack of specificity in this rhetoric—beyond generally raising awareness through information and expression and feeling.

As artists and activists look back over the cultural turn through the lens of climate crisis, they are beginning to see things differently. There is a growing sense that, as long as certain political commitments to imagination prevail—commitments based on open-endedness and ambiguity as much as awareness and affect—"we push off the work of organized collective action to another moment." The politics of imagination may not be an alternative to capitalist realism. The politics of imagination may be one of capitalist realism's most cunning forms.[6]

What if capital doesn't actually make it difficult, let alone impossible, to imagine alternative economies or other worlds beyond climate crisis? What if imagination is exactly what capital offers, what it prefers, because such acts pose almost no threat to capital accumulation or economic reproduction? What if this politics of imagination not only makes sense to us but also thrives in corporatized, middle-class art worlds and academies because it is consigned to arrangements that already exist? What if the environmental arts and humanities do not really offer political critiques of climate crisis? What if they are precisely the situation of politics that the capitalist climate crisis is itself rendering sensible and aesthetic?

If these were simply questions of whether, in making tomorrow, we should proceed in the manner of architects or bumblebees—whether we should raise our world in our imaginations before raising it on the ground (or vice versa)—then the matter would be academic. But it's not. It is a matter of political strategy.

As the cultural turn continues taking shape in the rearview mirror, one thing seems clear for the road ahead. In the same way that levels of awareness and the severity of the climate crisis are inversely proportional, the crisis is continuing apace regardless of all our wonderful acts of imagination. Awareness has not saved us. Neither will imagination.

15

A Behavioral Change Is Gonna Come

Like the culturalists, psychologists also believe music has a power that can be harnessed to link climate awareness with climate action. Unlike some culturalists, though, they are not satisfied with imagination.

The influence of musicians who write songs about climate crisis, who adopt environmentally friendly practices, and who use their platforms to raise awareness—all this "is rarely measured empirically," says one music psychologist. "The mechanisms through which music might help us to address the climate crisis are not examined." By specifying those mechanisms through the experimental rigor of psychological research, "we have the potential to use our knowledge of the power of music to try to change individuals' behavior in small ways that collectively address climate change and ultimately, save lives."[1]

Although not dealing with music as such, the wider field of environmental psychology has been hard at work on similarly behavioral questions for some time. The main issue here is straightforward. How can people become more environmentally friendly? One of the field's foremost answers is to engineer environmentally friendly behavior according to a logic of stimuli and responses that resemble the context of a game.

Gamification is a way of turning nongame settings, like acts of everyday citizenship, into situations resembling sports or arcades. Gamification makes everything about scoring points, gaining rewards, leveling up, competing—and sometimes getting a pep talk or a nudge when your activity becomes dissatisfactory or delinquent.

Duolingo, the most downloaded language app, offers an illustrative example. This Pittsburgh-based educational technology company has gamified language learning, thanks in no small measure to its rootedness in the psychology of behavioral modification. When you are playing the game and doing well, the app gushes praise and showers you with treasure. Not coincidentally, environmental psychologists trying to promote "pro-environmental behaviors" have found success through forms of positive reinforcement that turn the hard work of climate action into "fun, noticeable, and rewarding opportunities" to make "cool choices for sustainability."[2]

Positive reinforcement has an effect on behavior. But gamification knows an even stronger motivator. Guilt. If your study habits start sliding, for example, Duolingo will let you know. The app's infamous nudges range from passive-aggressive reminders ("learning Norwegian requires daily practice") to the manipulative pleas of its teary-eyed owl mascot ("we haven't seen you in a while"; "you've let Duo down"). Likewise, environmental psychologists have found that, more than positive reinforcement or feelings of pride in one's sustainability efforts, it is guilt that most consistently "motivates pro-environmental outcomes."[3]

There are already numerous "gamified apps for sustainable consumption" out there in the world. In music, one of the most visible examples is offered by Coldplay. The band's World Tour App was originally developed for their 2022 voyage and was used through their 2024 and 2025 tours supporting the album *Moon Music* (made from recycled plastic bottles). Among numerous features, from news and videos to concert dates and augmented reality, the app also includes a few simple mobile games.

These games promote the band's sustainability initiatives, like their famous off-grid electricity. In addition to being provided by kinetic dance floors, event power is also generated as concertgoers ride stationary bicycles while taking in the Coldplay spectacle. The related promotional in-app game asks you to solve puzzles (essentially mazes) and then to frantically pedal an on-screen bike in order to "feel the electricity of being one with the show." This is certainly one way that music has gamified sustainability. But the real pro-environmental behavior here, the real link between awareness and action, comes from the app's more general gamification of consumption.

Echoing a widespread consensus in music, which says that live music is the industry's biggest polluter and audience travel is the biggest problem in that category, one of the app's main purposes is to help you as a music fan "make low impact travel choices, share your carbon report, and get a merch discount if you travel green." An earlier generation of music fandom, which held up the live concert as more real than the recorded artifact, was about the authenticity of being there. True fandom today is becoming about the authenticity of getting there.

Hot air balloon is one way that Coldplay suggests traveling to their concerts, although the option is not available from, say, Oslo to Melbourne, two cities I have called home. Flight seems like the most sensible option. But the app glows disapproving red. The trip involves over 30,000 kilometers of travel, 3,000 kilograms of carbon, and thirteen newly planted trees to offset the damage. "Your mode of travel is a high emitter of carbon," the app says. "Consider greener modes of transport."

Public transit is offered as a separate option, even though it still involves flight, and most air travel is collective transportation anyway. Unsurprisingly, the option throws up nearly the same figures as before. Yet now the app glows approving green. "Your mode of travel is a low/zero emitter of carbon. You are a carbon hero!" As a reward, the promised discount code appears: 10 percent off Coldplay merchandise. The discrepancy between flight and public transportation would be hard to explain if not for the fine print of the app's end user license agreement. Coldplay's software developer, SAP, "does not warrant the meaningfulness of transportation methods."

Whether through the use of happy emojis or sad ones, red glows or green ones, and whether their meaningfulness is warranted or not, these things seem to work.

There are established criticisms of such apps. For example, the games often reinforce sustainable behavior with coupons that encourage further consumption, which seems self-defeating. Gamified environmentalism also represents a version of the responsibilized, incremental theory of social change that, notwithstanding its veneer of psychological legitimacy, never seems to add up in the way its advocates intend.[4]

There is also something unsettling about the resemblance between trying to solve the climate crisis by using a simple stimulus, like gamified music fandom, and efforts to program how people think, how they feel, and what they do. It feels not just reductive but also downright manipulative to manufacture human responses based on animalistic conditioning. That kind of thing may be okay for rats and dogs, but not us. Critics have long dismissed aspects of this thinking as behaviorist.[5]

Behaviorism belongs to the bygone, like color TV and transatlantic accents. It is not something many contemporary psychologists of music or the environment would subscribe to. But maybe they should. Radical behavioral science is actually not about minds, feelings, or stimulus-response conditioning. Radical behaviorism is about changing the circumstances of the world—developing technologies of behavior that facilitate cultural engineering—in ways that bring the remote consequences of our actions (like climate crisis) closer to the meaningful core of decision making. Behaviorism is about changing what is outside of people, not what is inside them.[6]

Say what you will about behaviorism. The standpoint has certainly been used to support some weird arguments. For example, one of behaviorism's most famous proponents wrote that the pursuit of values like human freedom and dignity actually interferes with something like our bare survival. But at least behaviorism encourages reflection on first principles. From this perspective, the trouble is not that environmental psychology and app development are behavioral rather than social sciences. It is that they are not behaviorist enough.

The real issue is not whether it is possible to empirically verify that music works to alter climate behavior. It obviously does. The issue is that reducing sustainable behavior to a fun (or guilt-ridden) game is to smother larger political questions about climate crisis. Gamification may nudge users toward certain individual choices of sympathetic consumption. It may also reward them for those choices. Yet there are no personal solutions to social problems. Sustainability games are unlikely to ask users to question what is worth sustaining in the first place or to reflect on problems before running after solutions. Games encouraging pro-environmental behavior do not see that the true crisis of the climate is abstract, not concrete. They are unlikely to nudge users toward class politics or to reward them for organizing against capital.

Music can surely affect our behavior. But is this the role we want for culture in the struggle for the planet's future? A role where political considerations are subordinated to evidentiary questions about whether something works or not?

16
No Place

Music moves us. It sets us in motion. This is undoubtedly true. (Not all truth is produced in a lab.) But where does it take us? The expectation that music we happen to like sends us somewhere we should want to go rests on a prior assumption. It supposes that music hasn't itself already been set in motion by other social forces.

Björk offers one way of illustrating this point. The impossibly imaginative musician is well known for her environmental advocacy. She is a favorite example, not least among researchers, of someone who engages productively—progressively—with environmental themes in her music and imagery.

The title track on Björk's 2017 album *Utopia*, to take just one example from her catalog, is rich in sounds of bird and breath and flute and word, cryptic but suggestive, while the video has an amniotic ambiance, all pink and blurry, nourishing a land of pixies and other fantastic creatures. It is a world purified of toxicity, she sings, as the overall aesthetic works to "redistribute the categories of nature, culture, human, and animal." Through all this, Björk molds a paradise in song—a "posthumanist" audiotopia where "human and animal music sound together," pointing to "new ways of existing in our environments."[1]

"Utopia" is a beautiful, moving track. The appeal of hearing it as a posthumanist hymn is apparent. Throughout history, the designation of humanity has been used to justify all sorts of monstrosity. This includes recklessly elevating human societies above the natural world. The designation has also been used to wrongfully elevate some people above others. Many of humanism's deeds have been a disgrace.

This explains why it is common today to hear artists and authors calling for a more level playing field. Calling for shifts away from humanism, toward posthumanism. Away from anthropocentrism, toward ecocentrism and multinaturalism. Away from human exceptionalism, toward multispecies salons and parliaments of things. Away from subjects and their symbols, toward objects and their materialities. Away from stable human being, toward more-than-human becoming.

In all this, social order is no longer understood to be based simply in the interactions of intentional individuals. It is seen as always emerging, jazzlike, in contingent networks of humans *and* nonhumans. (Laying emphasis on the conjunction is a ritual whose forward-thinking credentials require neither explanation nor verification.) Insofar as Björk's audiotopia exemplifies the best of these ideas, insofar as it points beyond the monstrosities of humanism, we had better listen up.

But it is possible to wonder what else this song might be setting in motion, owing to the ways the music has itself been set in motion. Maybe the philosophy of posthumanism, for example, is not all it's cracked up to be. When it comes to our economic arrangement, there is reason to believe that whereas posthumanism sees itself as a rebellion against the logics of capital and western rationality and colonial modernity, it may actually be an expression or fulfilment of those logics.

Something similar was once said about postmodernism and late capitalism. The cultural logic of a given moment does tend to symbolize and secure the social organization of its economy.

As capital continues to rob this planet and its people, increasingly under cover of greenness, posthumanism's leveling and redistribution of the capacity for social action can seem less about prosecuting the offense than providing its alibi. Capital wants nothing more than to debase human agency. It loves to play shell games with accountability. It is glad for public life to shatter itself into billions of tiny entanglements and experiences. It needs us to accept that the world is beyond our comprehension and out of our control.

If the posthuman drives of capital can happily coincide with various celebrations of posthumanism, then the philosophy starts to look suspiciously like an accomplice in the shifting regime of accumulation. Our goal should be to evaluate the politics of the ideas we use to make sense of the world and what makes those ideas both possible and logical, not to simply use them in descriptions of their own appearances.[2]

It is also worth asking about the new modes of existence Björk is inviting us into. What kind of world is it where to be purified of toxicity is to be purified of the everyday infrastructures that are required to purify toxicity? In other words, where does the poop go?

Sewers are nowhere to be found in Björk's audiotopia. The same goes for all infrastructures, not just of sanitation but also of transportation, communication, cultivation—really, all large-scale modern construction projects. Björk not only presents a small-is-beautiful environmental philosophy. She also portrays reverence for a detached fable of nature as

Still taken from Björk's "Utopia," 2017. Creative direction by Björk, directed by Warren Du Preez and Nick Thornton Jones. Used with permission.

a timeless, harmonious wilderness that appears both post-human and prehistoric.

Such visions of nature have only ever been made possible through various infrastructures of life and labor in the city and the country—and they typically have only been made available to those who can afford the experience of detachment. As historians of nature have understood, "we mistake ourselves when we suppose that wilderness can be the solution to our culture's problematic relationships with the nonhuman world, for wilderness is itself no small part of the problem."[3]

The video's parting shot is most eyebrow-raising in these terms. As the camera pans out and we see Björk's world in full for the first time, listeners are presented with a sanctuary floating alone in the ether. It is a reference to Thomas More's original island utopia, certainly. It may also suggest something about the musician's feelings toward her native Iceland. Either way, this is an appealing, comforting vision—a soft-focus sigh of relief. However, the very characteristics of purity and detachment that offer the relief also align this vision with some of the more flagrant examples of the inequality wrought by capital today. Björk's audiotopia looks like a suburb or a seastead. It resembles a gated community.

Gated communities—whether the real but figurative islands of suburbia or the imaginary but literal islands of seasteading—are part of a social reality where capital has long carved out spatial divisions that reinforce inequality and where free-market fundamentalists are increasingly molding spaces where capital can slip the cuffs of democratic governance. Alongside tax havens, special economic zones, and much else, this is a world where it has also recently "become routine to frame enclave building projects in terms of climate." Like Björk's audiotopia, these islands for the fortuned few are increasingly "pitched as models of 'sustainability,' responsible solutions to sea-level rise, and paragons of a 'low-carbon' lifestyle." Forget about the enclosures that enable these enclaves, or the lives that enable these lifestyles. Just about the only thing sustained in this fantasy of sustainability will be the war with otherness that fuels the accumulation of capital and flanks the environmentalism of the rich.[4]

If Björk's track is an invitation to a particular vision of the future, it is a future where public splendor will be subordinated to the ongoing privatization of luxury. It is a future that will keep pulling the wool over its own eyes when it comes to the planetary factory needed to keep that world running. It is a future where much of what passes for theory and critique will sound a lot like the situation it is meant to be theorizing and critiquing. It is a future perfectly consonant with the ongoing transition from fossil capital to green capital.

How much should we make of the sport of interpretation? Quite a lot, I think. Consider a parallel example from a classic 1990s cartoon, *Captain Planet and the Planeteers*. The premise of the series is that Gaia, voiced in early seasons by Whoopi Goldberg, "can no longer stand the terrible destruction plaguing our planet." So she calls on five children from Africa, North America, Eastern Europe, Asia, and South America, who team up with Captain Planet to save the world by being part of the solution rather than the pollution.

It may seem straightforwardly positive to present children with a form of entertainment that also educates them about saving the planet. However, the program has been shown to perpetuate exactly the kind of leftist ecoliberalism that promotes consumption as the path to conservation and offers simplistic individual actions as the answers to complex social problems. The cartoon, presumably despite the good faith of its creators, portrays a "contradictory environmental ethic that seeks at once to improve the environment and maintain existing social relations." It also exemplifies a "liberal misdirection of children's very real willingness to help save the planet."[5]

In the same way that children's superheroes like Captain Planet can sneak a standpat ecoliberalism through the back door of edutainment, grownups' superheroes like Björk can carry that ecoliberal common sense right through the front door of the avant-garde.

It is not simply that Björk's "Utopia" affords different interpretations. The point is to watch out for the easy fit of interpretation with the interpreted, of theory with the theorized. It is to know that what things seem to mean may be an effect of what they are unable to say, and that acts of explanation are sometimes precisely what needs to be explained.

If artifacts and interpretations exist somewhere between their articulation of a social consciousness and a political unconscious, then the issue is not just that awareness and imagination won't save us. It is that the wrong awareness can set the wrong things in motion. The wrong awareness can make for the wrong action. It can lead us astray.

A major irony of the social sciences and humanities is that critical, symptomatic readings like the one I have attempted here have, in the twenty-first century, given way to an intellectual paradigm less concerned with what things mean than with whatever they might make possible. Such work is presented as a liberating move beyond a fusty tradition founded on texts, readings, and pessimism. Yet this recent work has exactly followed the rise of a media environment where the truth content of our worldly representations has never been less stable. At the same time, our ability to do things with those representations, our ability to experience the full range of what they make possible, has never been less hindered. Which is a hindrance itself.[6]

In such a scenario, it is not just that cultural artifacts are congruent (intentionally or not) with the conditions of their production. It is that thought and scholarship too have an uncomfortably comfortable relationship with their economic arrangement and social architecture.

Far from an anomaly or slip, this is actually an enduring feature of the academy. We have already seen the example of the so-called cultural turn, where the highly structured and totalizing logic of capital since the 1970s was accompanied by intellectual teardowns of ideas like structure and totality. It is also the case that, since the very beginning of professional social science, the academy has always been "a major instrument for developing and sustaining liberal ideology," despite robing itself in critiques of that ideology.[7]

The same holds true today. "It's a mistake," says one critic, "to believe that the humanities departments of our universities are hotbeds of leftism, even though both the supposedly left-wing professors and their right-wing critics love to believe it. In fact, they're more like the research and development division of neoliberalism."[8]

It would be unfeeling to suggest that music's recomposition is simply the R&D division of ecoliberalism. There is a lot of promising environmental music out there, in Björk and beyond. There is also plenty of good environmental work being pursued in music's degree paths and research programs. Ecological music matters. The same goes for broad movements in environmental arts and humanities. Still, it is undeniable that the rise of all this, since the early 2000s, precisely parallels the wider music world's recomposition.[9]

More often than not, these approaches take up the same problems and advocate the same solutions as the recomposition itself. I know I have. I've talked about raising awareness, generating affect, and setting things in motion. I've trusted in the power of imagination. I've traced supply chains and followed waste streams. I've tallied carbon footprints. I've turned phrases where it seemed like simply mentioning materiality or posthumanism, or stressing humans *and* nonhumans, automatically conferred worthwhile insights and progressive politics onto whatever I was writing.

Now I'm not so sure.

17
The Future, Wouldn't That Be Nice?

Ancient Greek philosophers knew it. Ancient Chinese emperors knew it too. So did North African theologians at the beginning of the common era. The same goes for pretty much every other metaphysician, politician, and aesthetician down through the ages.

Poets say something like it about verse. Dramatists have a similar idea about theater. Writers of all kinds say it about art, from anarchists to fascists, journalists to humorists, litterateurs to fictioneers. Teachers say it about the classroom, revelers say it about the dance floor, puppeteers say it about the spotlight. Julie's Bicycle says it about culture in general. Some philosophers say it about Björk in particular. It is repeated endlessly in relation to music.

The expressions are many, but the idea is simple. Culture and the arts—music especially—can set people in right relation to themselves and others in the body politic (never mind the cosmos). Opinions vary on the principles of those relations and that body. But the belief that music contributes to social order, that it may model social order, that changes in one are functions of changes in the other—all this has been accepted for a long time. It has been embraced by many people in many places.

A related idea comes in regular and extra-strength varieties. For some, cultural and artistic works are understood as doing more than configuring social order or modeling it. Culture and art are vaunted as vanguards of social change, because people who make artifacts tend to be especially alert to the drift of history. It's not that such works can themselves change the world, they say. It is that these artifacts can change people, who can change the world.

For others, taking stronger doses, cultural forms and performances are not just early registrars of shifting social sands. They are said to be revelators of wider changes to come. They are said to anticipate or foretell those changes. Making music one way in the here and now foreshadows how the world could be another way in the there and then. Music may even give shape to that world, because sound travels faster than the speed of life.[1]

There is a name for this set of ideas, for this will to fashion the social formats of tomorrow in the cultural practices of today, for this philosophy of the future. It goes beyond a doctrine of configuration. It is known as a politics of prefiguration.

The history of the future is long and varied. But prefigurative politics has taken a distinct and influential shape since the cultural turn of the long, global 1960s. As a species of imaginative politics, prefiguration made sense to a disaffected generation that preferred freeing their minds to getting bossed around by authorities. As a political strategy, it appealed to social movements frustrated by top-down and state-led initiatives. It spoke to people tired of the usual relationship between political means and social ends. They were excited by the possibility that the means of political activity could themselves exemplify or equivalence the ends of social order they hoped to achieve.

Prefiguration has been expressed in innumerable movements and manifestos. Its modern musical flashpoint, though, can be found in two books. Both were published in 1977. One was written in French, by a neoliberal economist. The other was written in English, by an unorthodox music educator.

In *Noise: The Political Economy of Music*, French economist Jacques Attali suggested that musical forms did not simply register or reflect the conditions of their production. Rather, he said, "music is prophesy." It has an ability to herald, in the cultural arena, changes in the economic arrangement. This view was softened as the work was taken up in the Anglosphere following its 1985 translation. Music was understood less "to presage new forms of socioeconomic organization" than "to foreshadow changes in sensibility, ideology, and structures of feeling." But it was prefigurative nonetheless.[2]

Our unorthodox educator, Christopher Small, presented something similar in his book *Music, Society, Education*. The subtitle says it all: *A Radical Examination of the Prophetic Function of Music*. Drawing on theories of art, culture, and imagination going back to the nineteenth century, Small suggested that these realms "can reveal to us new modes of perception and feeling" and "can make us aware of possibilities of alternative societies whose existence is not yet." Music, especially, "outlines the forms of that potential society which lies still beyond our grasp."[3]

The forms of prefiguration outlined by the economist and the educator have had a lasting influence on how musicians invest their vocation with political meaning and energy. Musical forms and textures as well as the performance dynamics of classrooms, ensembles, concerts, and genres—especially experimental music, improvised music, and networked music—all apparently serve as scale models for future civic engagements and forms of social cohesion. Particularly significant here today is the idea that fading ideals of postwar social democracy may still be found in music and sharpened through it.

Many musicians hold these truths to be self-evident. Some register degrees of skepticism about how musical means might achieve these social ends. Either way, it is not necessary to linger on the question of whether music really works in the ways profiled by prefigurative politics. To embark on that corroborative quest would be to miss the point. It would be to sociologize the evidentiary one-eyedness of some psychology and most app development.

Neither, though, is it necessary to go in the opposite direction, searching for ways to falsify the prefiguration hypothesis. The idea that the arts and music are telling signs of social order and social unrest may take different shapes through history and culture. It certainly has a particular shape in our moment. But the basic idea persists across times and places. It is clearly something we want to believe in.[4]

Rather than treating prefiguration as a fact to be confirmed or a myth to be busted, it is better to recognize the anthropological role of prefiguration in the structure of our mythology. With myth, as with fetishism and metrification, it is not only veracity that matters. Just as the best test of proof is occasionally to make a meal of the pudding, sometimes the best concretion of a truth is to absorb a story in its telling.

Myths tend to be about creation—of the universe, of customs, and so on. Yet mythology is often "creativity turned against itself." That is, while "mythic thought is endlessly creative," mythology places its moments of creation in the mist-enveloped past, originating outside human timelines and social orders, fixing people's relationships to one another and the world around them in a form of primordial authority. Myths may be about creativity, but they function conservatively.[5]

The myth of prefiguration is about nothing if not creativity. Prefiguration says not that present social orders were created and fixed in some misty past, but rather that creativity is abundant, the present is plastic, and the future is open. When it comes to tomorrow, the quiet secret of the world is that it is something we make, and we could just as easily wake up one day and decide to do things differently. The choice is ours. To begin the transformation, an imaginative cultural form like music would be as good a place to start as any.[6]

On the surface, prefiguration seems to run against the usual track of myth. Look closely, though, and the overarching tendency of mythology becomes apparent. Prefiguration can be a Trojan horse of the status quo.

Consider that the recent history of prefiguration matches the economic, philosophical, and sociological shifts of the cultural turn, where white collars and mental work overtook blue collars and manual work in both population proportion and political prominence. It flatters this class to insist that their imaginative effort, not the physical labor of various others, is what really matters in making the world—and changing it. It suits them to believe that their cultural creations are the prelusive seeds from which otherwise elusive future social orders will grow. This view of politics unites immediate interests (imaginative acts pay the bills or at least accrue cultural capital) with fundamental interests (imaginative acts both reveal the current social order and render the way beyond it). As with sympathetic consumption and production, prefiguration can ordain a kind of upside-down marriage of convenience, full of progressive feeling but without much practical effect.

All mythology involves the projection of desires, fantasies, uncertainties, and anxieties. In cases where prefiguration places philosophical idealism and individual voluntarism before historical materialism and collective organization, the myth risks amounting to little more than a projection of well-intentioned professionals searching for meaningful models of cultural production and consumption. The risk, in other words, is that the middle class has simply championed a politics that embodies the truth it wants to see. The trouble—the reason prefiguration's symbolic displays can underwrite the social reality they pretend to displace—is that capital is famously adept at engendering opposition in its own image.[7]

That is one way of explaining the myth's appeal. Consider also that prefiguration matches a history in which culture, the arts, the humanities, and the social sciences have been plunged into austerity and precarity, managerialism and instrumentalism. Consider too that this history was crystallizing at the same time our economist and educator published their books. It's no accident.

In such settings, arguments for arts subsidy and research funding are often obliged to state their cases in terms of benefits to economy, society, psychology, and, increasingly, the climate. It makes strategic sense here for artists and academics to lean into the enticing myth that music has a special, prefigurative potency in those realms. If free jazz models the fates of liberty and democracy, if live music predicts the fortunes of a city's nightlife and tourism industries, if top-forty tunes forecast the future of sustainability (or at least offer a way of coping with a climate-changed existence), then surely culture must be studied and supported. Surely states and businesses will open their wallets.

None of this reflects the neoliberal indoctrination of anyone. These are understandable responses to unfavorable conditions. They are simply attempts to get by in circumstances most people did not choose.

In fact, prefiguration seems to thrive in settings where imaginative possibility is all some people have left—or where people have little else to begin with. There can be no doubt that a variety of speculative futurisms and radical traditions, since at least the twentieth century, have offered not only hope and maps but also vocabularies and forms that have helped envision somewhere in advance of nowhere. But to the extent that idealism and materialism still cut a line between reactionary and progressive politics, prefiguration is the cultural knife that capital is only too happy for us to bring to its structural gunfight.[8]

There is a politics of newness that inhabits prefiguration as a philosophy of the future. This opens onto another way that managerial values can inform cultural ideals about how we make tomorrow. The operative phrase here is that music is the canary in the coal mine when it comes to the future of the cultural industries.

Beginning in the nineteenth century, canaries were used as carbon monoxide detectors in mining operations. Because their little avian lungs and birdy metabolisms would succumb to the poisonous gas before humans were fatally affected, canaries offered a useful early warning system in the mines. Music was not the first media industry or cultural arena to commandeer the songbird analogy. Authors offered canary-in-the-coal-mine theories of the arts during the 1960s. Video games threatened budget movie rentals in the same terms during the early 1990s. But the canary imagery did become firmly attached to music in the year 2000. It was then that the

US National Academies published a report called *The Digital Dilemma: Intellectual Property in the Information Age*.[9]

The general idea of the report, which is also reflected in scholarship, is that music prophesies the political economy's social order in ways other than those suggested by the economist and the educator. More than other cultural forms, music here is understood to be "at the forefront of the turbulent changes to the production, distribution, and reception of culture galvanized by digitization." Partly this means that music serves as an early warning sign of the hazards that digitization has posed to existing business models and ways of earning a living—whether file sharing three decades ago or artificial intelligence today. But it has always simultaneously meant that music serves as an effective means of sniffing out new markets—new technology, new creativity, new value, new growth.[10]

The Digital Dilemma is explicit about this. Although the report is obviously swept up in a particular set of concerns in the history of online commerce and copyright, its authors also calmly itemize some instructive earlier episodes in the interlocked histories of technological development and industrial opportunity. Lending libraries did not end bookstores. Photocopying did not end publishing. Videotapes did not end cinema. In fact, "in each case, the new development produced a new market far larger than the impact it had on the existing market." Neither, then, was the internet likely to end the record industry. "Some suggest that the ability to download music will increase sales by providing easy purchase and delivery twenty-four hours a day, opening up new marketing opportunities and new niches." Here music appears less like a sentinel species than a divining rod.[11]

No wonder prefiguration is so alluring, not just to musicians but also to institutions like funding agencies and academies.

Like the managerial and instrumental situation that defines arts subsidy, many of these bodies are nowadays invested in market-friendly novelty and innovation. Conveniently, music and culture are ahead of the curve, revealing those opportunities sooner than other industries, and researchers have the skills to help point the way.

Something similar can be said for recent government reports promising that participating in cultural forms like music has a "clear knock-on impact" that can "strengthen democracy." This sounds good. Democratic renewal is preferable to the alternative storms on the horizon. Yet it is worth underlining that the form of postwar social democracy idealized in such reports, and in a lot of music research, has always been engineered to underwrite the future of capital. To preform the renewal of that social democracy in music culture is, on some level, to perform the endurance of capitalist market relations.[12]

Shortly before he died in 2011, our unorthodox music educator asked a group of students: "So what about music education and climate change?" We know how he would have answered himself.[13]

"A musical performance," wrote the educator, "brings into existence relationships that model in metaphoric form those which they would like to see in the wider society of their everyday lives." To change a musical performance—by, say, imagining and enacting a posthuman audiotopia—would be to prefigure different social relationships. It would be to begin remaking tomorrow and, in so doing, to help solve the climate crisis.[14]

Of course, this is appealing. It's not just that this kind of thing receives accolades in the art world and funding in the academy. It is something we want to believe. It is a matter of the truth we want to pursue. Prefiguration is part of the structure of our mythology. But it is also a species of imaginative politics that, on some level, is a bill of goods sold to us by capital. In this sense, the real enemy of prefiguration is neither those who would falsify it nor those who would factify it. Often enough, prefiguration's true adversary is its own devotees.

So, what about music and climate change? It is possible for us to give an answer to this question that is different from the educator's answer. If music somehow communicates the future, it does so in a voice that simultaneously calls out for new markets, announces new product cycles, and sometimes whispers hints about new forms of political economy. Music does so as a function of a broader social transformation in which market principles and middle-class proclivities risk being centered as the most meaningful dimensions of political life, and where the logic of capital is often affirmed in its effacement. It does so in concert with a politics of culture where making tomorrow means capitalizing on the future.

Counterintuitively, then, if prefiguration is a manifestation of the future in some present arrangement, this can serve not to bring that future closer but instead to hold it back, to delay its arrival, to enact the postponement that "brings capital into being and allows it to multiply." Music appears consonant with a broader economic arrangement that, through many shifts in how it has accumulated and reproduced itself, has always used smokescreens of optimism and excitement to obscure the truth of its dishonesties and damages. It has always promised "a better tomorrow, tomorrow."[15]

Figuring out how to stop living on that credit, how to stop living off this deferral of the future—which is what the shift from fossil capital to green capital really means—is crucial. Doing so would certainly help to solve the appearances of the climate crisis. But it would do so as an effect of rebuking and rebuilding more essential economic arrangements and social architectures. And the way to achieve that is only partly to assert that we make tomorrow. The larger question is about how we make tomorrow history.

18
One Thing to Another

We of this generation stand midway between two eras. When we look backward, we see our past like a great tidal wave that is now receding but that was magnificent indeed in the sweep of its socially purposeless power. When we look ahead, we see something new and strange, something undreamed of in our various philosophies. What we see ahead is the threat of dissolution. The great wave piled up too much wreckage—of nature, of obsolete social patterns and institutions, of human blood and nerve.

The above words are not mine. They are adapted from a pamphlet published nearly a century ago, in 1932, called *Culture and the Crisis*. Although the document addressed itself to the economic situation surrounding the Great Depression, its pages might as well have been written for the Great Recomposition. It speaks uncannily to our own crisis, our own midway point between two eras (fossil capital and green capital), our own inescapable conclusion that the climate crisis is being managed by those who produced it.

In 1932 the pamphleteers were appealing to "writers, artists, scientists, teachers, engineers, to all honest professional workers" to gather themselves and stand against capital.

They were not making an argument about the role of culture in raising awareness through artifacts and aesthetics. Nor were they stating culture's case in the imaginative distribution of affective sensibilities. Naturally, they were doing those things in their own cultural works. But their main point was about discharging the potential political energy that would come from joining the cultural workforce to the industrial workforce.

The pamphlet's authors had "no faintest desire to exaggerate either our talents or our influence." Their aim was to recognize shared fundamental interests across classes, despite more immediate yet divisive interests, and their goal was to "strike hands" in an "alliance with the only militant force which seeks renovation." They understood that the choice of predominantly middle-class cultural workers, then as now, was "between serving either as the cultural lieutenants of the capitalist class or as allies and fellow travelers of the working class."[1]

Our choice may be the same, but our time is different. Cultural work is not the same as it used to be. The industrialization and massification of culture in the twentieth century may be recognizable to us today, as our precursors. Yet the contemporary conditions of commerce and uncertainty in that industry would have been hard for them to imagine back then.

Industrial work is not the same either. The hands of those with whom cultural workers wished to strike have changed. Organized labor unions of yesteryear have been disaggregated. Their movements have been halted. Meanwhile, the archetype of labor in those days, the tough-guy white factory worker, is less central than in the past. He hasn't disappeared. He still matters. But he has partly given way to a globally distributed workforce. That workforce includes not only traditional sites of manual labor. It crisscrosses with teaching, healthcare, warehouse picking and packing, the unpaid labors of social reproduction, and more.

Our time clearly requires its own strategies and forms of organization. Romance and nostalgia have no place here. Yet culture does, just not in the way half a century of culturalism might lead us to expect.

Today the basic message of *Culture and the Crisis* is more important than ever. For all the cultural turn's advances, its limitations are clearer now. In addition to leftist critiques of cultural politics, similar limitations apply to the forms of dissent that went along with them—which include, from 2010 to 2020, from Hong Kong to Chile, "more mass protest than at any previous point in human history, exceeding the famous global cycle of contention in the 1960s." Again, in both the 1960s and the 2010s there were specific successes. On balance, though, "nowhere did things turn out as planned." And "in far too many cases, things got much worse." With cultural politics it is hard to escape the conclusion that, despite the uprisings, things have gone downhill.[2]

This returns us to Adolph Reed Jr. Reed has never had time for cultural politics. He definitely doesn't think much of music's political potential. Cultural politics, Reed says, is pretense. Playacting. It is worse than no politics at all.

Reed's words are strong. We may want to temper some of what he says. We may want to reach for counterexamples. But it is impossible to look back over the history of the cultural turn and deny his point entirely. This is especially true considering the relationship between the ascent of widespread cultural awareness and imaginative engagements with climate issues, on the one hand, and the decline of the climate's actual condition, on the other. Still, when Reed describes what his lifelong experiences in activism, organizing, and research suggest is necessary to build the kind of political movement that could tackle capital head-on, it is telling that, in a way, he begins with culture.

For most people in most places, Reed says, the basic frame of reference, the common social denominator, "is the employment relation, the fact of working." He continues: "A politics focused on bringing people together around such concerns cannot be built, especially not in its early stages, mainly through big events. It grows much more from one-on-one interaction and with small groups of coworkers, neighbors, friends, and other associates. Part of the intellectual work of this kind of organizing is cultivation of an ongoing discussion, linked to practical political activity, around making collective sense of how the social order and its mystifications are reproduced on a daily basis."[3]

The role of culture that Reed is describing in the US context comes up again and again in the work of seasoned activists and organizers, drawing lessons from India to Poland and Sri Lanka to El Salvador. One of those activists, Marta Harnacker, writing on Latin America, says the art of progressive politics "is not a matter of advancing the most radical slogans, or of carrying out the most radical actions—which only a few join in because they scare off most people." To her list we might add composing the most radical, prefigurative music or conjuring the most radical, posthuman imaginations.[4]

This is not to say that certain people will necessarily be repelled by certain things. We have a generation of research showing that aesthetic preferences do not always map neatly onto social positions. (There are plenty of medics who like mashups and mechanics who like Machaut.) But it is to suggest that what matters most here are the basic building blocks of culture. Being radical, Harnacker emphasizes, is not primarily about slogans and actions. Instead, it involves "creating spaces where broad sectors can come together and struggle. Understanding that we are many and are fighting for the same objectives is what makes us strong and radicalizes us." This, for her, as for Reed, is where politics and culture really meet.[5]

There are significant obstacles to advancing such political cultures in the globalized twenty-first century. Another veteran activist observes one such hindrance. "For many unionists in India," Rohini Hensman says, "First World workers constitute a labor aristocracy whose prosperity has been built on imperialism. On the other side, there is often palpable rage against Third World workers on the part of workers in developed countries, who believe that the passivity and docility of the former are responsible for their own loss of jobs." Regardless of how well the labor aristocracy thesis or the scapegoating of immigrants explain our reality (they don't), these ideas are responsible for some of the most divisive feelings that prevent us from seeing the fundamental interests we all share, from seeing how many we really are.[6]

In the face of such animosities and obstacles, the best of cultural politics responds: "To the command of neoliberal globalization—'workers of the world, compete'—we must answer with that old slogan of the global justice movement: 'workers of the world, unite.'" But that time-tested motto needs updated forms of cultural expression. "We need to put it in new words, new songs, new figures," says one advocate. This will allow us to gather a "yet unimagined, unrepresented collectivity."[7]

Surely this is part of the story. At the same time, the responses of activists like Harnacker are more along the lines of Reed. She shows that obstacles have been overcome simply through "face-to-face contact" and "a chance to meet." The authors of *Culture and the Crisis*, meanwhile, saw hope for the future in coalitions across the class divide—a progressive cultural front that took seriously the potential of both cultural works and cultural work.

It didn't pan out. But that doesn't mean it couldn't. History, they say, moves more by failure than success.

Alison Tickell is right. Culture needs to be part of the climate conversation. But the understanding of culture advocated by Julie's Bicycle, which permeates our general climate common sense as well as the specific recomposition of the music world, has to be understood according to its historical and political parameters.

The point is not to invert the inversion of the cultural turn. But having faced this turn honestly and learned its lessons, it is time to admit that it is probably not the imaginative cultural work and cultural works of musicians and artists, or intellectuals and students, or managers and administrators, or individual sympathetic consumers and entrepreneurial sympathetic producers, or middle-class professionals of various sorts who will be the principal agents in the changes that are required not to solve the climate crisis but instead to make inroads against capital and class. Not on their own, anyway. The potential of this class falls in striking hands with another.

If new progressive cultural fronts and class coalitions are to be built from the ground up, they will depend on musicians and artists telling new stories about the world and envisioning new tomorrows. But coalition is not simply a matter of mythology or imagination. If new fronts and coalitions are to be built, they will also depend on the self-organization of subcontracted and precarious cultural industry workers, on joining tiers of musicians, artists, academics, writers, and journalists with the part-timers, the casual workers, the bearers of social reproduction, and others in restaurants and garment factories and warehouses and places of education and care across nations and around the world. Cultural workers are significantly responsible for the sounds, pictures, and stories by which we make sense of ourselves and our situations. Cultural workers and their cultural works can make us stop and think again. They can make us feel things. They can move us. They may even tell us something about the future. But after half a century of privileging a particular version of this mythology—of privileging the politics of culture's artifactual, imaginative, and affective dimensions in a particular way—it seems worth placing at least as much emphasis elsewhere.[8]

Such an emphasis would begin with people talking about this stuff on the job and in the community, contributing to protests and actions and rallies, walking picket lines, joining unions (and their choirs). It would be about being present in all those small moments where cultural politics means simply getting together—and, sometimes, belting a chant, banging a drum, having a dance, or sharing a song.

Shifting emphasis in this way is obviously not about giving up on music. Nor is it about giving up on culturalism and returning to economism. The goal is not to go back to anything. The goal is to move forward. It is to recognize that the cultural turn of the past half century has been accompanied by a counterrevolution—a violent and intentional process of class decomposition—and that our job is to help recompose that class, to help recompose class politics, to help recompose class power. One day historians may even look back on a movement from the cultural turn to the great recomposition. Music will be vital there.

We of this generation may buy all the right stuff, invent all the right tech, impose all the right regs, tap all the right energy. We may harness all the power of music, raise all the awareness in the world, and channel all the imaginative force in the universe, proving in our psychologies and knowing in our mythologies that music culture is the missing link between climate awareness and climate action—that this is how we will make tomorrow.

All of this is needed, and none of it requires a confrontation with capital. None of it asks anyone to break the constitutive tensions between economic priorities and ecological principles, immediate and fundamental interests, or resistance and resignation that constrain our actions, maintain the class divide, and sustain the unsustainable. None of it goes where we truly need it to—not unless it passes through a class-based confrontation with capital.

Without that crucial step toward class recomposition and majority coalition, without that affirmative yet additive political gesture, music's own recomposition, along with everything else in mainstream climate politics, will be stuck with half measures and partial answers instead of well-formed questions and definitive actions. Technical, institutional, and cultural solutions to climate crisis will continue to appear—more and more, faster and faster, in music and beyond. But despite all our best efforts and intentions, and regardless of every desperately needed reform, such responses will threaten the very dissolution they set out to prevent. Without better problems, the wreckage will keep piling up. And the world will keep breaking down, sweeping with magnificence but without social purpose, from one thing to another, wave after crashing wave.

Postscript

Recomposed began with a set of questions. How do we make music more sustainable? How can music help solve the climate crisis? What do we do? This is where my attempt to answer those questions has ended up.

I am mindful that in some ways I have simply collected and recollected things that many people already understand when it comes to the possibilities and limitations of green capital, of reform in relation to sterner stuff, of what music might contribute to social stability and social change.

I also know that every "and" I have added to a "yes" is easier said than done. Writing a book is not the same as the guts it takes for people in music to pursue technical, institutional, or cultural solutions to the climate crisis.

It is clear too that I have mostly turned out a compass, a way of locating a direction. Convincing people to move in that direction, figuring out how to organize the passage, deciding where to end up—this is where the real work begins. And I am aware that there are countless people who already devote their lives to that work, to the politics I only truly fell into as I wrote this book. Compared to them, I'm a futon revolutionist. But all change begins somehow, and somewhere love and justice shine.

Since starting this book in 2019 and publishing it in 2026, I'm not sure how much has changed—though a lot has happened. Sickness, fire, war, tariffs, AI, the persistence of inequality. The persistence of resistance.

In all this, some say we have entered a whole new political economic order. Others say it's business as usual. Some say things are better than ever. Others speak of enshittification.

One certainty is that there have been backlashes against a variety of more and less progressive causes. This includes many of the broad paths to solving the climate crisis described in these pages. Environmental advisements and protections of all sorts are being ignored, delayed, dismantled, abandoned. Reading the news today, it is hard to miss a shrugging surrender to the fact that critical temperature targets will be overshot. The transition from fossil capital to green capital has apparently slowed or stopped. Maybe it never even started.

Whether this situation deepens our resignation to it, whether it leads us to double down on climate solutionism, whether it prompts us to disavow ecoliberalism as an enemy ecology and dig into more radical problems—all this would alter the meaning of our moment. And it would make the core political message of this book more true, not less.[1]

Either way, the Great Recomposition is still happening.

Some of its exponents are gaining momentum. EarthPercent is raising more and more money. ClimateEQ has further broadened its scope into "high quality multi-industry consultancy and training." Heidi Lenffer is hard at work, and FEAT has expanded into "financing conservation and biodiversity," which it now considers to be "a dual mandate" alongside its clean energy mission.[2]

For others, traction is trickier. Take Green Vinyl Records. The company went quiet for a while. But Harm Theunisse says everything will be okay. Green Vinyl is now "powering" a project called Good Neighbor Music, "an ecofriendly LP record manufacturing company." Good Neighbor boasts Erykah Badu as a spokesperson, offers colorful plastic in shades of ocean floor and salsa verde, and has been involved in releases ranging from Finneas to Shellac to the Lafayette Afro Rock Band.

Evolution Music's bioplastic record is still in a phase of research and development. They are ironing out new supplier relationships and tinkering with their plastic compound, having recently adjusted the Evovinyl recipe to include polyhydroxyalkanoates. "The results have been astonishing," Marc Carey says. Evolution is conducting additional tests at large pressing facilities in Europe and the United States, talking about building a manufacturing plant of their own, and they held a launch event for their signature product in late 2025. Exciting as these developments are, the company remains on the starting line of full commercial viability. Nevertheless, what Steve Charter said is right. It is happening. It's just that "happening" is hardly the simple matter we often wish it was.

Out in Vancouver, Billy Bones is still pressing records. He's still making music too. In fact, the Vicious Cycles just put out a new album, *Get Wrecked*, and I'm in town for the release party.

Billy meets me on a corner of Commercial Drive, and we head north to Lynn Canyon. The album release is later tonight, but the band is in shape, so he's made time for a short hike across a suspension bridge, springtime sunshine slicing through the conifers.

As I jump in the car, Billy jumps into how things are at Clampdown. Six years in, and it's going alright. But making records is not an easy buck. Whenever the plant is up and running smoothly, that virtually guarantees something is about to go wrong. If it isn't one thing, it's another.

These days, sourcing plastic is not the problem it was a few years back. Billy has plenty of that. His whole team is on the job too, and the machines are happily stamping away. Then, on cue, the chiller breaks down.

Pressing vinyl happens in a molten moment. Cooling and curing that record takes longer, in a process that basically involves sitting around in controlled heat and humidity. This is admittedly less spectacular than the high-temperature hydraulics it takes to squash a glob of plastic into a disc. But cooldown is just as crucial to producing a quality product. It is also just as intricate as any other aspect of making records. There are systems of glycol tubing, roof-mounted fan units, and much more.

Cooling records correctly prevents them from disfiguring in ways that can create playback problems. It prevents records from getting warped into shapes like saddles and dishes. Nobody wants dishy vinyl.

Through the canyon, we talk about the guitars we built or didn't. About older amps, like Garnets, and newer bands, like Amyl and the Sniffers. About motorbikes. About the election. About life.

One thing that doesn't come up so effortlessly is the environment. I have to ask. Clampdown no longer brands itself as one of the greenest pressing plants in the world. Billy hasn't commercially made records out of potatoes or windows or anything other than plain old polyvinyl chloride. He's heard about developments in greener discs, of course. And his skepticism is healthy.

Billy knows that peddling products as environmentally friendly tends to involve at least some snake oil. Ingenuously or otherwise, there is always a chance the proposed green remedy will contribute to the same problem it is being sold to solve. As long as the economy works the way it does, he says, as long as we're all locked in competition, as long as there is no release valve for the environment, it only ends one way.

But Billy Bones is not bleak. He is still exploring alternatives. He is still doing whatever he can. In fact, he has two bags of nonpetroleum PVC kicking around his shop, good for about 200 records. The bags do not contain Evolution Music's plant-based compound, Evovinyl. Nor do they contain the non-PVC substance of Green Vinyl and Good Neighbor. These bags are of another kind. This plastic is a variation of biovinyl, which is being piloted commercially by stars like Billie Eilish.

At the end of the day, biovinyl is still PVC. So it's not problem-free. But instead of petroleum derivatives it uses things like corn, sugarcane, wood, straw, and non–food grade vegetable oil in the early stages of its polymerization. This is supposed to reduce the footprint of each record by 95 percent. It also raises the wholesale cost by about a dollar a disc.

Getting people interested in biovinyl is not difficult, Billy says. It's harder, though, to get buy-in and follow-through, so for now he's going to use the bags himself. It may not be the window-frame solution he was looking into when I first visited Clampdown. Nor is it the thrift store road trip he had imagined. But Billy Bones does plan to make a few of the new Vicious Cycles records in this way.

Get Wrecked is an album of "rip-roaring party music." It is not just a record you can dance to, Billy says. It's a record you can't not dance to.[3]

As the Vicious Cycles take the stage for their hometown release party, in a banquet hall behind an Irish bar, as they rip and roar into the first single off the new album, "Hold On Tight," the packed crowd gets involved, no one can not dance, and things come full circle. Amid the likelihood of climate overshoot and the stutter steps of recomposition, in all these tight spots where economic priorities rival ecological principles, where immediate interests throttle more fundamental ones, we're back to that once neglected topic of social inquiry. Fun.

Live, "Hold On Tight" is fun as anything. The video also has a good thing going, with its gorilla suits and skeletons and eight-bit, stop-motion charm. Lyrically too—*wrap your arms around me*, *never let me go*, *I'm in a dream when you're riding with me*—this is a straightforward track about a simple joy: "riding with your favorite person on the back of your motorcycle." Scratch that surface, though, and you hit the base mettle of this concept band: "motorcycles as allegory over chrome plated punk."[4]

The record is full of singalong self-reflection and anthemic daydreaming that also nods to a bigger picture. Songs about naked beach debauchery are also about "the tiny lives we're expected to lead" under capital. Catchy forty-seven-second blasts offer advice on "dealing with the conformist naysayers of the world." And "dour castigations of fiscal inequality and political strife" are simultaneously "such a toe-tapping good time." Even the cover art captures the band's modus vivendi. "A tuff looking scrub on a minibike says a lot about who we are."

These are the kinds of opposing forces that give torque and motion to the symbolic economy that defines the Vicious

Cycles (and plenty of music) as much as the political economy that defines Clampdown (and the recomposition). Between appearances and essences, between what is and what might be, principle and priority, life and work—this is where the action is, the potential in the actual, where the rubber meets the road. And the history of all this, like the history of the world itself, is the history of struggle, class struggle, the struggle for fun.[5]

"Hold On Tight" is about simple joy, real fun. It is also, Billy says, a metaphor for life.

Stick together.

Stick it out.

"We're gonna make it."

I am not optimistic. But I hope he's right.

Acknowledgments

Friends and colleagues have considered many versions of this book. It is not possible to do justice to their influence—or my gratitude—through citation alone.

Over a decade ago, Tom Everrett gifted me this whole research direction. He has since prevented it from going off the rails, more than once.

In recent years, Tami Gadir shared many resources she knew I should know about. When I finally got it, the world looked different.

For reading drafts of this book at various stages, I would like to thank Kelly Burdick, Eric Drott, Ellis Jones, Polly Leger, Clinton Morin, Anne Pasek, Jonathan Sterne, Marek Susdorf, Nicholas Tochka, and Tom Western. For consulting on thoughts and questions throughout the process, thanks to Georgina Born, Alexandrine Boudreault-Fournier, Matt Brennan, Ainslie Coghill, Nicola Dibben, Christopher Faulkner, Simon Frith, Ieva Gudaitytė, Kai Arne Hansen, Fabian Holt, Dean Jenkinson, Emil Kraugerud, Chang Liu, Sadie Menicanin, Franny Nudelman, Meredith Pal, Marianna Ritchey, Elodie Roy, John Shepherd, Tore Størvold, Catherine Strong, and Paul Théberge. Even when faced with badly underdeveloped material, everyone responded with uncommon ministry and grace. I only wish I could give back as good as I've got.

Parts of the book were written while I was visiting the Department of Art History and Communication Studies at McGill University. Thanks again to the late Jonathan Sterne, who made things happen, gave pivotal input, and lent me a guitar. It was also very good to spend time with (and learn things from) Carrie Rentschler, Darin Barney, and Will Straw. Michèle Paquette and Matthew Hunter made the visit a logistical breeze.

Other parts of the book were written while I was a Macgeorge Fellow at the University of Melbourne's Conservatorium of Music. I am grateful to the Macgeorge Bequest for its support. Another thank-you to Nicholas Tochka, and to Michelle Frencham, for making it all come together. Nick had a big influence on this project. I would additionally like to say how much I have appreciated the wider academic and activist communities in Melbourne over the years, including Gay Breyley, Shelley Brunt, Paul Long, Benjamin Morgan, Kat Nelligan, Ian Rogers, Sam Whiting, the Music Industry Research Collective, XR Westside, Solidarity—plus, again, Tami Gadir and Catherine Strong.

(I was set to spend time developing this project as a guest in the School of Music and Creative Media Production at Massey University. Travel restrictions meant that I never made it to New Zealand. Nonetheless, I am grateful to Catherine Hoad for the invitation and for the work she put into our application.)

Many more have shared invaluable ideas, resources, and support on various levels and in various settings. A thousand thanks to all my coworkers and students at the University of Oslo in the Department of Musicology, the Oslo School of Environmental Humanities, and the Unruliness research network. Thanks as well to the Low-Carbon Research Methods Group and my new coworkers at the University of Winnipeg.

Some ideas in this book have appeared previously, in different forms. I have also tested these ideas in several presentations. I'm grateful to all my editors, to everyone who made those events possible, and to all those audiences for their input.

Thanks as always to friends and family both near and far.

There is another group of people who I am very glad to acknowledge. These are the individuals directly involved in music's recomposition.

Thanks to everyone who offered comments, questions, correspondence, and coverage related to my previous book. Such interactions number in the hundreds and stretch across the world. The interest and curiosity of these musicians was the spark, and is the core, of this book.

Thanks, too, to a variety of industry and industry adjacent individuals. They equally made this book possible. In no order: Billy Bones at Clampdown Record Pressing; Harm Theunisse and Pieter van Ettro at Green Vinyl; Larry Jaffee, Sarah Murray, my fellow panelists, and everyone else at Making Vinyl in Hollywood and Offenbach; Steve Charter, Marc Carey, and everyone at Evolution Music; Sarah Ditty, Vicky le Lerre, Joel Gardner, Golchehr Ghavami, Sarah Williams, Wayne Snow, and everyone at EarthPercent; Adam Gardner and Carson Byrum at Reverb; James Dove, Anthony Daly, and my classmates at ClimateEQ; Heidi Lenffer at FEAT; Tim Shiel at Spirit Level; Marte Wulff at Spillerommet; Simon Lee at ReRoot; Subin Kim at Kpop4planet; Nick Jackman and Mike Walsh at Serenade; Fay Milton and Lewis Jamison at Music Declares Emergency; Darren Henderson and everyone at the Earthfest Music Summit; Ruben Planting; Tim Anderson and Reyna Bryan at Good Neighbor; Martin Charter at the University for the Creative Arts; Trevor Davis of Trevor Davis and Associates (formerly of IBM); Jen Cregar at Terra Lumina Consulting; Ben Swanson at the Secretly Group; Andy Inglis at KRS Open; Rob van Wegen at Eurosonic Noorderslag; Solveig Korum from Arts and Culture Norway; Sharmin Sultana Sumi; Ian Stanton at Beggars Group; Roxy Erickson at Music Climate Pact; Pål Bråtelund at Vilvit as well as his associates: Pete Downton (former deputy CEO of 7digital), Mike Jbara (CEO of MQA), Ty Robert (former CTO of Universal Music Group), Howie Singer (former CTO of Warner Music Group), and Bob Stuart (founder of MQA).

In the interest of disclosure and honesty, to help readers as they form opinions about the arguments in the book, I will mention a few things about my relationship to these individuals and their efforts. Some connections are uncomplicated. Most people simply wanted to talk about their work, and I was simply glad to learn about it. Our working relationships are, I hope, mutually beneficial in that it is good for them to be in this book and good for the book to have them in it.

A few of these relationships are more involved. At EarthPercent, the music industry climate nonprofit, I am currently the lead expert for greening music, having served previously on the expert advisory panel. In addition to sometimes acting as ambassadors for the organization, advisory panel members review proposals to make use of EarthPercent's funds. They then make recommendations, which go to the organization's trustees, who ultimately decide where the money goes. Lead experts perform a similar service, with a bit more responsibility in the sequence of evaluation. Advisory panel members are bound by a confidentiality agreement, which I signed for a two-year term, and they are offered an annual honorarium of £500, which I declined. I later signed a similar agreement for a three-year term as a lead expert and was offered a yearly honorarium of £750, which I similarly declined.

With support from Coldplay, Live Nation, and the Warner Music Group, I worked with Hope Solutions and the Massachusetts Institute of Technology's Climate Machine initiative to help produce their assessment report on live music and climate change. As a member of the advisory committee, my role was to review drafts of the report and to offer comments about how best to understand and reduce the environmental consequences of the concert industries. The work was voluntary.

I am also on the advisory board of Evolution Music, the company that has introduced a bioplastic LP. Board members are meant to help the company achieve its goals. To this end, I have offered my experience in meetings of various kinds and contributed to panels at industry events like Making Vinyl. I have also provided soundbites for use in promotional materials and made introductions to people interested in Evolution's product. For their services, board members are offered 125 shares in the company. If the shares begin to accrue significant value, I will place them in a kind of anonymous trust that will administer the assets until a later date. The idea is to mitigate a potential conflict of interest.

Conflict of interest, though? I hope all these projects—and more—will succeed. That is no secret. Still, I do not think my enthusiasm on this front, or my involvement in it, has clouded my judgment in seeing the bigger stage on which music's recomposition is playing out. My perspective should be evident throughout the book.

This perspective is of a particular kind. It is not about fault finding or finger wagging or applecart upsetting. But neither is it about taking the world at its word, simply gathering and describing various facets of lived experience. Rather, an important goal of social inquiry is to end up with explanations of "the possible discovery-production of patterns of determination."[1]

I have tried to explain the problem-solution matrix that music has constructed for itself—the Great Recomposition—along such lines. I've described the recomposition as a fact of life that has been shaped by the material parameters of a social space in which people are encouraged to pursue solutions to a problem that has been defined for them in terms of its appearances (climate and crisis) rather than its essences (capital and class). I have tried to fasten this phenomenon to its economic arrangement and social architecture in order to open up possibilities, not just for better solutions, but for better problems. I have done so while also asking after the patterns of determination that have made possible my own discovery-production of patterns of determination.

In other words, to talk about social inquiry is to talk about how to construct an object of study and how to navigate a researcher's relationship to that object of study. We are talking about how to respect the common sense of the world, and learn from it, while also producing some kind of break with that common sense, while also trying to understand the conditions that would enable both the common sense itself and the ability to break with it, while also attempting, somehow, to hold it all together—to put it all into conversation.

There are numerous, longstanding dialogs and debates about this kind of inquiry. There are also many fancy ways of describing it. But the basic operation is not so complicated.

It is to initiate "a process of mutual education in which both sides work together to explore possibilities that are outside lived experience."[2]

Knowledge and politics are always inside lived experience, are always states of being in the world. They are also always, at their best, something more.

A final word on how I did this research, and on what method means for the conditional character of knowledge.

Regarding my headquarters in a wealthy country known for its role in perpetuating fossil capital, which afforded me the two things I needed to put together this book about the climate crisis—time for research and writing, budget for travel and books—I do recognize the ironies. I hope I have been able to avoid the worst potential hypocrisies.

From this perspective, some readers may wonder about a book on the condition of the climate that has racked up its share of frequent flyer miles. Others may wonder about this book's existence as something made of ink and paper. Should I not have conducted all my research online? Should I not have published digital-only, as a small file at a low resolution? Should I not have outlined this book's carbon footprint in a methodological appendix?

There is nothing wrong with authors (not to mention musicians) choosing to fly less. Nor is there anything wrong with inquiring into the carbon emissions and resource intensities of writing and research. Yet my goal in *Recomposed* has been to warrant some reflection about racing toward carbon footprints, individual responsibilities, and consumption activities as the go-to ways of understanding and trying to improve the situation of the world (and our relationship to it).

When it comes to research design, the book is actually unmethodical. It all just sort of—happened. Although I am confident that I have identified, described, and explained something about a general phenomenon in the world of music, one that echoes a larger social moment of climate adaptation and mitigation, the people in this book (as well as those who constitute its backdrop) are simply the ones I had the good fortune to meet in the wake of my previous book about the environmental history of the record industry. Most of the time, I got to know these people because they happened to contact me. Or we met by chance at an event of some kind. Good fortune, to be sure. But good fortune is not the same as dumb luck.

Who I met in my research is a function of the patterns of a world that was allowed to reveal itself to me. I did not try to intervene in that process, or I did so minimally. This means that there are very real ways in which this book is limited by forms of citizenship, geography, and culture whose coordinates most resemble my own. Saying so is not to suggest that diversities of many kinds do not exist in the space of recomposition. They do. The point is to understand something about the kinds of paths authors take, and the kinds of encounters that are made possible, when we follow roads that are paved for us in advance.

At the same time, the patterns of climate thought and action that I encountered in my research were consistent across the musicians I met from South Africa, South America, South Asia, and many other places. The social explanation for such consistencies obviously cannot be citizenship, geography, or culture. The left and right versions of climate solutionism that are the targets of this book's materialist inquiry do not obey such categories, even as they are inflected by them.

Recomposed is about capitalist class relations of domination between the North, or the economic core, and the South, or the economic periphery, as well as between upper and lower classes in the core and periphery themselves. Yet the book is written from within, and with a focus on, the centers of economic power. It is also written by someone, and mainly for others, who inhabit the contradictory location in our globalized social architecture known as the middle class. This is partly the result of a methodological accident, albeit a predictable one in that it occurred at the intersection of pre-paved roads. It would be possible to write a very different book—a more methodical one, certainly, but also a book that would seek out and lift up firsthand accounts of ecological suffering and persistence in those places that the economic core exploits, oppresses, and makes peripheral.

Yet I have allowed my methodological accident to take place. And I have allowed my focus to remain on the centers of economic power, because I believe that understanding capital in its latest and leading forms—in its most twisted enormity—is essential to understanding how this arrangement affects the people and places that it peripheralizes. Such a focus also seems key to understanding the commonalities of subjection, and the possibilities of alliance, that both span the North–South divide and exist within its hemispheres. It seems key to building the kinds of class recompositions and majority coalitions that will be required to alter, not just of the appearances of the climate crisis, but the essences of its existence.[3]

Again, other projects are possible—and needed. But such is the project to which this book is committed.

Notes

Introduction

1. The plate described at the beginning of this introduction is made by a Polish company called Biotrem. See also Andrew Trendell, "Øya 2019 Is Inclusive, Green, and Feel-Good—Like All Festivals Could Be," *NME*, August 19, 2019.
2. Parts of this paragraph paraphrase the opening of Ivan Illich's *Energy and Equity* (Harper and Row, 1974 [1973]), 3. See also Jesse Goldstein, *Planetary Improvement: Cleantech Entrepreneurship and the Contradictions of Green Capitalism* (MIT Press, 2018); Nick Dyer-Witheford, "Struggles in the Planetary Factory: Class Composition and Global Warming," in Jan Jagodzinski, ed., *Interrogating the Anthropocene* (Palgrave Macmillan, 2018); Lucas Bessire, *Running Out: In Search of Water on the High Planes* (Princeton University Press, 2021); The Red Nation, *The Red Deal: Indigenous Action to Save Our Earth* (Common Notions, 2021); Thea Riofrancos, *Extraction: The Frontiers of Green Capitalism* (W. W. Norton, 2025).

3. I use words like *we* and *us* and *our* when writing about responses to climate crisis. There are good reasons to address readers by using plural pronouns—and good reasons not to. For background on my decision and the politics of universalism it implies, see Ellen Meiksins Wood, "Capitalism and Human Emancipation," *New Left Review* 167 (1988); Vivek Chibber, *Postcolonial Theory and the Specter of Capital* (Verso, 2013); Sylvia Wynter, "The Ceremony Found," in Jason Ambroise and Sabine Broeck, eds., *Black Knowledges / Black Struggles: Essays in Critical Epistemology* (Liverpool University Press, 2015); Laboria Cuboniks, *The Xenofeminist Manifesto: A Politics for Alienation* (Verso, 2018); Max Liboiron, *Pollution Is Colonialism* (Duke University Press, 2021); Jonathan Sterne, *Diminished Faculties: A Political Phenomenology of Impairment* (Duke University Press, 2021).
4. Another word for *mainstream* or *common sense*, in this context, would be *hegemonic*. See Kate Crehan, *Gramsci's Common Sense: Inequality and Its Narratives* (Duke University Press, 2016); Vivek Chibber, *The Class Matrix: Social Theory After the Cultural Turn* (Harvard University Press, 2022).
5. Make no mistake: capital will exploit oil reserves and invest in brown assets (as opposed to green ones) as long as it is profitable to do so. For background, see Nancy Fraser, *Cannibal Capitalism: How Our System Is Devouring Democracy, Care, and the Planet—and What We Can Do About It* (Verso, 2022); Max Ajl, "Theories of Political Ecology: Monopoly Capital Against People and the Planet," *Agrarian South: Journal of Political Economy* 12, no.1 (2023); Brett Christophers, *The Price Is Wrong: Why Capitalism Won't Save the Planet* (Verso, 2024).
6. I am paraphrasing a few sentences in this paragraph from Fraser, *Cannibal Capitalism*, 109. While I am emphasizing

the probability of a certain form of stasis in the passage from fossil-fueled capital to renewable energy capital, this is not to discount the possibility that the passage will eventually be understood as a crucial step forward in the development of some postcapitalist project.

7. Jem Aswad, "Coldplay's 'Music of the Spheres' Tour Drastically Reduces Band's Carbon Footprint, Sets New Standards in Sustainability," *Variety*, June 5, 2023.
8. For background, see Anwar Shaikh, "An Introduction the History of Crisis Theories," in Union for Radical Political Economics, ed., *US Capitalism in Crisis* (Monthly Review Press, 1978); Simon Clarke, *Marx's Theory of Crisis* (Macmillan, 1994); Janet Roitman, *Anti-Crisis* (Duke University Press, 2014).
9. For background, see Robert Brenner and Mark Glick, "The Regulation Approach: Theory and History," *New Left Review* 188 (1991); Imre Szeman, "How to Know About Oil: Energy Epistemologies and Political Futures," *Journal of Canadian Studies* 47, no. 3 (2013); Andreas Malm, *Fossil Capital: The Rise of Steam Power and the Roots of Global Warming* (Verso, 2016); (s)Connessioni precarie, "Seven Theses on Climate Change and the Ecological Regime of Accumulation," *Transitional Social Strike*, October 21, 2021; Éric Pineault, *A Social Ecology of Capital* (Pluto, 2023).
10. The middle class is a mess. See Marlene Dixon, "What's in a Name? A Critical Review of Barbara and John Ehrenreich's Professional-Managerial Class," *Synthesis* 2, no. 3 (1978); Erik Olin Wright, *Class, Crisis, and the State* (New Left Books, 1978); Ricardo López and Barbara Weinstein, eds., *The Making of the Middle Class: Toward a Transnational History* (Duke University Press, 2012); Barbara Ehrenreich and John Ehrenreich, *Death of a Yuppie Dream: The Rise and Fall of the Professional Managerial Class* (Rosa

Luxemburg Stiftung, 2013); Hadas Weiss, *We Have Never Been Middle Class: How Social Mobility Misleads Us* (Verso, 2019); Göran Therborn, "Dreams and Nightmares of the World's Middle Classes," *New Left Review* 124 (2020); David Roediger, *The Sinking Middle Class: A Political History of Debt, Misery, and the Drift to the Right* (Haymarket, 2022).

11. Quote and examples from Tad Skotnicki, *The Sympathetic Consumer: Moral Critique in Capitalist Culture* (Stanford University Press, 2021), 3. See also Lucia Hulsether, *Capitalist Humanitarianism* (Duke University Press, 2023).
12. Quotes from Lawrence Glickman, *Buying Power: A History of Consumer Activism in America* (University of Chicago Press, 2009), 2. See also Skotnicki, *Sympathetic Consumer*, 4–6; Erima Dall, "Boycotts and the Fight for Social Change," *Solidarity*, April 23, 2014; Ishay Landa, "The Negation of Abnegation: Marx on Consumption," *Historical Materialism* 26, no. 1 (2018).
13. See Chibber, *Class Matrix*, and Søren Mau, *Mute Compulsion: A Marxist Theory of the Economic Power of Capital* (Verso, 2023).
14. For background, see Steven Bernstein, *The Compromise of Liberal Environmentalism* (Columbia University Press, 2001); Ted Steinberg, "Can Capitalism Save the Planet? On the Origins of Green Liberalism," *Radical History Review* 107 (2010).
15. Matteo Battistoni, *Middle Class: An Intellectual History Through Social Sciences* (Brill, 2022), xii; Catherine Liu, *Virtue Hoarders: The Case Against the Professional Managerial Class* (University of Minnesota Press, 2021), 77.
16. For background, see Fred Rose, *Coalitions Across the Class Divide* (Cornell University Press, 2000); Robert Johnston, *The Radical Middle Class: Populist Democracy and the Question of Capitalism in Progressive Era Portland, Oregon* (Princeton University Press, 2003); Murat Arsel, "Climate

Change and Class Conflict in the Anthropocene," *Journal of Peasant Studies* 50, no. 1 (2023); Stefania Barca, *Workers of the Earth: Labour, Ecology, and Reproduction in the Age of Climate Change* (Pluto, 2024).

17. There are various ways of understanding the overlaps of music, capital, class, and environment in relation to social change. For a few recent sources that have influenced my approach—which attempts to be neither long on capital but short on class nor long on class but short on capital, and which tries to avoid being long on resistance, protest, and mobilization but short on vision, direction, and organization—see Archie Green et al., eds., *The Big Red Songbook* (PM Press, 2016); Anna Bull, *Class, Control, and Classical Music* (Oxford University Press, 2019); Marianna Ritchey, *Composing Capital: Classical Music in the Neoliberal Era* (University of Chicago Press, 2019); Aaron Allen and Jeff Todd Titon, eds., *Sounds, Ecologies, Musics* (Oxford University Press, 2023); Eric Drott, *Streaming Music, Streaming Capital* (Duke University Press, 2024); François Ribac, Isabelle Moindrot, and Nicolas Donin, eds., *Music and the Performing Arts in the Anthropocene* (Routledge, 2025); John Street et al., *Our Subversive Voice: The History and Politics of English Protest Songs, 1600–2020* (McGill-Queen's University Press, 2025); Noriko Manabe and Eric Drott, eds., *The Oxford Handbook of Protest Music* (Oxford University Press, forthcoming); Anna Morcom and Timothy Taylor, eds., *The Oxford Handbook of Economic Ethnomusicology* (Oxford University Press, forthcoming). See also Matt Brennan's Doughnut Music Lab (University of Glasgow).
18. I am paraphrasing and adapting Chibber, *Class Matrix*, 153.
19. See Matthew Huber, *Climate Change as Class War: Building Socialism on a Warming Planet* (Verso, 2022), 281. I am pointing to the potential of unions, not their perfection, as

means of organizing the politics of labor as well as making improvements within capital and perhaps moving beyond it.

20. Max Ajl, *A People's Green New Deal* (Pluto, 2021), 4, 8–9. Part of my framing here is taken from Christopher Faulkner, "In Defence of a Library," *Asylum* 34, no. 4 (2016).

1. The Ecological Record

1. "I've had people who want us to put real blood in records," Dunster says, "which we won't do." See Nate Jackson, "One of a Kind: Erika Brings Specialty Vinyl to the Masses," *OC Weekly* 22, no. 3 (2016). Additional detail in this section is compiled from other reportage and my own interactions with Dunster in 2019.
2. Quoted in Gary Graff, "Erika Records Founder Looks Back on 40 Years of Pressing Vinyl," *Billboard*, August 9, 2021.
3. Ben Homewood, "Beggars Group and Ninja Tune Unveil Plans to Go Carbon Negative," *Music Week*, April 19, 2021.
4. The comparison between pop videos and popping corn is adapted from Will Pickett et al., *DIMPACT: Methodology* (Carnstone, 2022), 30. The comparison to "physical" media comes from Kpop4planet.
5. The 2021 carbon estimate comes from Kpop4planet. See also Mark Sweney, "Universal Music Chief Predicts Billions of Dollars of Growth from Digital Listening," *Guardian*, September 21, 2021; Inder Phull, "What's Music's Next Format?," *Music Business Worldwide*, July 28, 2022.
6. The band is called the Dream. See Geir Rakvaag, "I en drøm (en vill en)," *Dagsavisen*, April 2, 2015.

2. Vicious Cycles

1. The review comes from *Classic Rock* magazine, while the comment on being a green pressing facility could once be found on the company's website, clampdownpressing.com.
2. Billy Bones quoted in Stuart Derdeyn, "Clampdown Record Pressing Inc. Pitches Tinder for Bands," *Vancouver Sun*, July 1, 2020.
3. On recycling generally, see Samantha MacBride, *Recycling Reconsidered: The Present Failure and Future Promise of Environmental Action in the United States* (MIT Press, 2012).
4. See Simon Frith, "Beyond the Dole Queue: The Politics of Punk," *Village Voice*, October 24, 1977; Gina Arnold, "Death in Vegas," in George McKay and Gina Arnold, eds., *The Oxford Handbook of Punk Rock* (Oxford University Press, 2025). See also Hanif Abdurraqib, "I Wasn't Brought Here, I Was Born: Surviving Punk Rock Long Enough to Find Afropunk," in *They Can't Kill Us Until They Kill Us* (Two Dollar Radio, 2017).
5. On fun, see Simon Frith, *Sound Effects: Youth, Leisure, and the Politics of Rock 'n' Roll* (Pantheon, 1981), 264ff.
6. Vicious Cycles, "Hot Dogs in the City," *Motorcycho* (Vicious Cycles MC / Pirates Press Records, 2019). Used with permission.

3. Green Vinyl

1. The whining sound, which other conference attendees also commented on, was probably not a fault in Green Vinyl's technology. Most likely, what I heard was something called chip squeal, an artifact of an early phase in the process of making records, which is out of Green Vinyl's control.

2. The quote in this paragraph circulates widely in the media and various reports. It can be traced back to Greenpeace reports from the early 2000s.
3. Quote from Bob Rolontz, "Revolution in Manufacturing—I: Injection and Compression Systems Double Record Output in 3 Years," *Billboard*, January 16, 1954, 1.
4. Quotes from Bob Rolontz, "Revolution in Manufacturing—II: Disk Makers Air Pro and Con of Injection Quality and Wear," *Billboard*, January 23, 1954, 13, 16; "Shelley Now Making Vinyl Records by Injection Mold," *Billboard*, July 21, 1962, 6.
5. Quote from Rolontz, "Revolution in Manufacturing—I," 36; "Revolution in Manufacturing—II," 16.
6. Quote from Jonathan Sterne, "Compression: A Loose History," in Lisa Parks and Nicole Starosielski, eds., *Signal Traffic: Critical Studies of Media Infrastructures* (University of Illinois Press, 2015), 39. See also John Durham Peters, *The Marvelous Clouds: Toward a Philosophy of Elemental Media* (University of Chicago Press, 2015); Alexandrine Boudreault-Fournier, *Aerial Imagination in Cuba: Stories from Above the Rooftops* (Routledge, 2021); Melody Jue, *Wild Blue Media: Thinking Through Seawater* (Duke University Press, 2020); Stefan Helmreich, *A Book of Waves* (Duke University Press, 2023); Nicole Starosielski, *Media Hot and Cold* (Duke University Press, 2021); Melody Jue and Rafico Ruiz, eds., *Saturation: An Elemental Politics* (Duke University Press, 2021).
7. I am paraphrasing and adapting Kirstin Munro and Chris O'Kane, "The Artisan Economy and the New Spirit of Capitalism," *Critical Sociology* 48, no. 1 (2022), 38. See also Richard Ocejo, *Masters of Craft: Old Jobs in the New Urban Economy* (Princeton University Press, 2017); Jason Pine, *The Alchemy of Meth: A Decomposition* (University of Minnesota Press, 2019).

8. The Smith reference is from *An Inquiry into the Nature and Causes of the Wealth of Nations*, vol. 1 (Methuen, 1904 [1776]), 16. The reference to the invisible hand having a green thumb comes toward the end of a speech made by Bill Clinton on the campaign trail during Earth Day, 1992.
9. Quote from Munro and O'Kane, "The Artisan Economy," 49.
10. See Juan Martínez-Alier, "The Environment as a Luxury Good or 'Too Poor to be Green'?," *Ecological Economics* 13 (1995); Peter Dauvergne, *Environmentalism of the Rich* (MIT Press, 2016).
11. See Imre Szeman, *On Petrocultures: Globalization, Culture, and Energy* (West Virginia University Press, 2019), 6; Petrocultures Research Group, *After Oil* (Petrocultures Research Group, 2016), 11–12.

4. Evolution Music

1. Although *Pitchfork* and *Billboard*, along with various social media posts, did announce this news on August 31, 2022, there were also earlier promotions in concert with Earth Day, in April 2022, as well as in July with Music Declares Emergency.
2. In addition to interviews and discussions, some of the descriptions and quotes in this section are drawn from the internal advisory board memos of Evolution Music.

5. Will It Scale?

1. Quote from Anna Tsing, "On Nonscalability: The Living World Is Not Amenable to Precision-Nested Scales,"

Common Knowledge 18, no. 3 (2012), 505. See also Alex Hanna and Tina Park, "Against Scale: Provocations and Resistances to Scale Thinking," CSCW Workshop, 2020.

2. Quotes from Tsing, "On Nonscalability," 506, 513; Anna Tsing, *The Mushroom at the End of the World: On the Possibility of Life in Capitalist Ruins* (Princeton University Press, 2015), 38. See also Katherine McKittrick, "Plantation Futures," *Small Axe* 17, no. 3 (2013); Janae Davis et al., "Anthropocene, Capitalocene, . . . Plantationocene?," *Geography Compass* 13, no. 5 (2019).
3. Elizabeth Abbott suggests that sugar has been the most devastating plantation crop. See *Sugar: A Bittersweet History* (Penguin, 2008).
4. Quote from Tsing, "On Nonscalability," 522.
5. Quote from ibid., 509. See also E. Summerson Carr and Michael Lempert, eds., *Scale: Discourse and Dimensions of Social Life* (University of California Press, 2016); Nick Seaver, "Care and Scale: Decorrelative Ethics in Algorithmic Recommendation," *Cultural Anthropology* 36, no. 3 (2021).
6. Quotes from Justin Joque, *Revolutionary Mathematics: Artificial Intelligence, Statistics, and the Logic of Capitalism* (Verso, 2022), 20, 22.
7. Quotes from Ryan Gunderson, "Problems with the Defetishization Thesis: Ethical Consumerism, Alternative Food Systems, and Commodity Fetishism," *Agricultural and Human Values* 31 (2014), 116.
8. Quote from Paul Dourish and Genevieve Bell, *Divining a Digital Future: Mess and Mythology in Ubiquitous Computing* (MIT Press, 2011). See also Nelly Oudshoorn, Els Rommes, and Marcelle Stienstra, "Configuring the User as Everybody: Gender and Design Cultures in Information and Communication Technologies," *Science, Technology, and Human Values* 29, no. 1 (2004); David Graeber, "Consumption,"

Current Anthropology 52, no. 4 (2011); Lucy Suchman, "Consuming Anthropology," in Andrew Barry and Georgina Born, eds., *Interdisciplinarity: Reconfigurations of the Social and Natural Sciences* (Routledge, 2013); Elizabeth Ellcessor, *Restricted Access: Media, Disability, and the Politics of Participation* (New York University Press, 2016).

9. Quote from Langdon Winner, "On Opening the Black Box and Finding It Empty: Social Constructivism and the Philosophy of Technology," *Science, Technology, and Human Values* 18, no. 3 (1993), 370. The final sentence of the paragraph paraphrases Alyssa Battistoni, "Latour's Metamorphosis," *Sidecar*, January 20, 2023.
10. Quote from Joque, *Revolutionary Mathematics*, 224.

6. Building Better Fetishes

1. Harvey Molotch, *Where Stuff Comes From* (Routledge, 2003), ix–x, 20.
2. William Pietz, "The Problem of the Fetish, I," *RES: Anthropology and Aesthetics* 13 (1985), 7–8. See also Edward LiPuma and Moishe Postone, "Gifts, Commodities, and the Encompassment of Others," *Critical Historical Studies* 7, no. 1 (2020).
3. For background, see Simon Frith, "The Industrialization of Music," in *Music for Pleasure: Essays in the Sociology of Pop* (Routledge, 1988); Jonathan Sterne: *The Audible Past: Cultural Origins of Sound Reproduction* (Duke University Press, 2003); Joshua Specht, *Red Meat Republic: A Hoof-to-Table History of How Beef Changed America* (Princeton University Press, 2019); Paris Marx, *Road to Nowhere: What Silicon Valley Gets Wrong About the Future of Transportation* (Verso, 2022).

4. Quote from Sebastian Pfotenhauer et al., "The Politics of Scaling," *Social Studies of Science* 52, no. 1 (2022), 21. On libraries and local streaming alternatives, see Liz Pelly, *Mood Machine: The Rise of Spotify and the Costs of the Perfect Playlist* (One Signal, 2025).
5. For background, see Hannah Krasikov, "The NFT Boom and Bust: Musicians as Productive Laborers in the Post-Streaming Music Industry," *Journal of Popular Music Studies* 34, no. 4 (2022); Ellie Rennie et al., *Developments in Web3 for the Creative Industries: A Research Report for the Australia Council for the Arts* (RMIT Blockchain Innovation Hub, 2022); Benjamin Morgan, Dave Carter, and Ian Rogers, "The Australian WEB3 Music 'Community' and the 'Indie' Mainstream," *DIY, Alternative Cultures, and Society* 3, no. 1 (2025).
6. Parts of this paragraph summarize and paraphrase Kate Crawford, *Atlas of AI: Power, Politics, and the Planetary Costs of Artificial Intelligence* (Yale University Press, 2021), 69; Matteo Pasquinelli, *The Eye of the Master: A Social History of Artificial Intelligence* (Verso, 2023); Emily Bender and Alex Hanna, *The AI Con: How to Fight Big Tech's Hype and Create the Future We Want* (Harper, 2025), 31, 13. See also Melissa Avdeeff, "Artificial Intelligence and Popular Music," *Arts* 8, no. 4 (2019); Georgina Born et al., *Artificial Intelligence, Music Recommendation, and the Curation of Culture* (Schwartz Reisman Institute for Technology and Society, 2021); Bob Sturm et al., "AI Music Studies: Preparing for the Coming Flood," *Proceedings of AI Music Creativity* (2024); David Hesmondhalgh and Raquel Campos Valverde, "Living With Music in the Digital Age," *European Journal of Cultural Studies* (2025); Pelly, *Mood Machine*.
7. See Wendy Woloson, *Crap: A History of Cheap Stuff in America* (University of Chicago Press, 2020). See also Sianne

Ngai, *Theory of the Gimmick: Aesthetic Judgment and Capitalist Form* (Harvard University Press, 2020).

8. Wyatt MacGaffey, "Fetishism Revisited: Kongo 'Nkisi' in Sociological Perspective," *Africa: Journal of the International Africa Institute* 47, no. 2 (1977), 172; J. Lorand Matory, *The Fetish Revisited: Marx, Freud, and the Gods Black People Make* (Duke University Press, 2018), xvi. See also David Graeber, "Fetishism as Social Creativity: Or, Fetishes Are Gods in the Process of Construction," *Anthropological Theory* 5, no. 4 (2005); William Pietz, *The Problem of the Fetish* (University of Chicago Press, 2022).
9. Rosalind Morris and Daniel Leonard, *The Returns of Fetishism: Charles de Brosses and the Afterlives of an Idea* (University of Chicago Press, 2017), xiii, viii.
10. Ibid., xiii. See also Matory, *Fetish Revisited*, 313.

7. The Carbon Question

1. "Carbon" in this book is a part-for-whole figure of speech that stands for greenhouse gas emissions and carbon dioxide equivalents more generally. Figures in this paragraph are based on websitecarbon.com. For a critical frame of reference regarding online emissions, see Anne Pasek, *How to Get Into Fights with Data Centers: Or, a Modest Proposal for Reframing the Climate Politics of ICT* (Trent University, 2023).
2. Quotes and paraphrases, both here and in some of what follows, are taken from the websites of the Carbon Literacy Project (carbonliteracy.com) and ClimateEQ (climate.eq.co.uk), as well as the slides from the ClimateEQ training.

8. ClimateEQ

1. On comic books, superheroes, and the structural anthropology of myth, see David Graeber, *The Utopia of Rules: On Technology, Stupidity, and the Secret Joys of Bureaucracy* (Melville House, 2015), 76–9, 207–27. On responsibilization, see Susanna Trnka and Catherine Trundle, *Competing Responsibilities: The Politics and Ethics of Contemporary Life* (Duke University Press, 2017).
2. For background, see Anita von Schnitzler, "Citizenship Prepaid: Water, Calculability, and Techno-Politics in South Africa," *Journal of Southern African Studies* 34, no. 4 (2008); Gavin Steingo, "Electronic Music and the Problem of Electricity," in Kyle Devine and Alexandrine Boudreault-Fournier, eds., *Audible Infrastructures: Music, Sound, Media* (Oxford University Press, 2021).
3. Quotes from Samantha MacBride, *Recycling Reconsidered: The Present Failure and Future Promise of Environmental Action in the United States* (MIT Press, 2012), 225.
4. See Hasok Chang, "The Myth of the Boiling Point," *Science Progress* 91, no. 3 (2008). See also Stuart Hall, "Culture, the Media, and the 'Ideological Effect,'" in James Curran, Michael Gurevitch, and Janet Woollacott, eds., *Mass Communication and Society* (Arnold, 1977).
5. The quotes in this paragraph and the general argument in this section are from Ken Alder, "A Revolution to Measure: The Political Economy of the Metric System in France," in M. Norton Wise, ed., *The Values of Precision* (Princeton University Press, 1995), 39, 42.
6. Ken Alder, *The Measure of All Things: The Seven-Year Odyssey and the Hidden Error That Transformed the World* (Free Press, 2002), 135.

9. Atmospheric Accounting

1. Quotes from John Fernández and Norhan Bayomi, eds., *Assessment Report of the Media and Entertainment Industry and Climate Change—Phase 1: Live Music, UK and US* (Massachusetts Institute of Technology, 2026), 56, 169–70.
2. Claire Stentiford, *Ecological Footprint and Carbon Audit of Radiohead North American Tours, 2003 and 2006* (Best Foot Forward, 2007), 3; Catherine Bottrill et al., *First Step: UK Music Industry Greenhouse Gas Emissions for 2007* (Julie's Bicycle, 2007), 5.
3. Cris Shore and Susan Wright, *Audit Culture: How Indicators and Rankings Are Reshaping the World* (Pluto, 2024), 35.
4. For quotes, info, and stats, see *IMPALA Carbon Calculator Report: Insights from Carbon Footprinting Independent Labels* (Independent Music Publishers and Labels Association, 2023), 30, 10.
5. The activities of these corporations precede but roughly parallel the announcement of the Music Climate Pact in December 2021, which is described further in the text. Similar collaborative initiatives include the Music Industry Climate Collective, announced in 2023, and the Music Sustainability Alliance, announced in 2024.
6. Quotes from *IMPALA Carbon Calculator Report*, 8; Spotify, *Equity and Impact Report 2022* (Spotify, 2023), 12. See also David Fickling and Elaine He, "The Biggest Polluters Are Hiding in Plain Sight," *Bloomberg*, October 1, 2020.
7. Quote from Jessica Green, *Rethinking Private Authority: Agents and Entrepreneurs in Global Environmental Governance* (Princeton University Press, 2014), 1, 139. See also Ferrial Adam et al., *Who's Holding Us Back? How*

Carbon-Intensive Industry Is Preventing Effective Climate Legislation (Greenpeace, 2011), 9, 51.

8. GZ Media, "Industry's First Life Cycle Assessment of a Vinyl Record," March 2025, gzmedia.com.
9. See Music Climate Pact and Vinyl Alliance, "Sustainable Supplier Program: Phase 1 Insights Report" (forthcoming).
10. Quote from Aaron Allen, "Ecoörganology: Toward the Ecological Study of Musical Instruments," in Aaron Allen and Jeff Todd Titon, eds., *Sounds, Ecologies, Musics* (Oxford University Press, 2023), 34. See also Anders Bjørn et al., "LCA History," in Michael Hauschild, Ralph Rosenbaum, and Stig Irving Olsen, eds., *Life Cycle Assessment: Theory and Practice* (Springer, 2017); Matthew Archer, *Unsustainable: Measurement, Reporting, and the Limits of Corporate Sustainability* (New York University Press, 2024).

10. Future Energy Artists

1. To clarify: the property destroyed in the 2013 New South Wales bushfires, one of nearly 200, was at the time the family home of Lenffer's partner. She now lives in the rebuild.
2. FEAT's main partners in the project were Future Super (a fossil fuel–free pension fund) and the Impact Investment Group (a former developer that now manages so-called ethical investment portfolios). In addition to my own interviews and meetings with Lenffer, and her blog on the FEAT website (feat.ltd), my account draws on press coverage surrounding the initiative in places like the Australian Broadcasting Corporation, the *Guardian*, and *PV Magazine*.
3. For a background here, see Rhys Williams, "This Shining Confluence of Magic and Technology: Solarpunk, Energy Imaginaries, and the Infrastructures of Solarity," *Open*

Library of Humanities 5, no. 1 (2019); Jordan Kinder, "Solar Infrastructure as Media of Resistance, or, Indigenous Solarities Against Settler Colonialism," *South Atlantic Quarterly* 120, no. 1 (2021); Imre Szeman and Darin Barney, "From Solar to Solarity," *South Atlantic Quarterly* 120, no. 1 (2021); Ayesha Vemuri and Darin Barney, eds., *Solarities: Seeking Energy Justice* (University of Minnesota Press, 2023).

4. The quote, as well as the language of solarity and promise, are from Vemuri and Barney, *Solarities*.
5. Quote from Petrocultures Research Group, *After Oil* (Petrocultures Research Group, 2016), 68. See also Eli Friedman, "U.S. Capitalism vs Chinese Capitalism: What Do We Need to Understand?," interview by David Camfield, *Victor's Children* (podcast), January 21, 2025.
6. Quote from Vemuri and Barney, *Solarities*, 22. See also Nicole Staroseilski, "Beyond the Sun: Embedded Solarities and Agricultural Practice," *South Atlantic Quarterly* 120, no. 1 (2021); Alyssa Battistoni, *Free Gifts: Capitalism and the Politics of Nature* (Princeton University Press, 2025).
7. The quote, which regards Peter Solterdijk, is from John Durham Peters, "The Media of Breathing," in Lenart Škof and Petri Berndtson, eds., *Atmospheres of Breathing* (State University of New York Press, 2018), 188.

11. EarthPercent

1. The 3 percent figure is not backed up by citations in the reportage I have seen. For Eno's figure, see Greg Cochrane, "Capitalism Didn't Understand Community: Brian Eno Steps Up the Climate Crisis Battle," *Guardian*, April 15, 2022.
2. Initial quotes from Adam Callan and Hiroki Shirasuka, "Opinion," in Fay Milton, ed., *Music Industry Climate Pack*

(Music Declares Emergency, 2021), 21. Later quotes from Tom Skinner, "Brian Eno Calls for 'a Revolution' in Music Industry's Approach to Climate Change," *NME*, October 12, 2021; Cochrane, "Capitalism Didn't Understand Community." Other parts of my description are compiled from one of Eno's meet-and-greets with the EarthPercent Expert Advisory Panel and his appearance on the *Sounds Like a Plan* podcast.

3. Quotes from the EarthPercent grant-giving strategy found on its website (earthpercent.org).
4. Quote from Heidi Lenffer, "Introducing the Solar Slice," FEAT blog, November 29, 2021, feat.ltd. See Sosefina Fuamoli, "No Mean Feat: How Aussie Artists are Saving the Environment," *Rolling Stone*, September 18, 2023.
5. Eno quoted in Skinner, "Brian Eno Calls for 'a Revolution.'"
6. See INCITE!, ed., *The Revolution Will Not Be Funded: Beyond the Non-Profit Industrial Complex* (Duke University Press, 2017).
7. Quote from Soniya Munshia and Craig Willse, foreword to ibid., xvi. See also Arundhati Roy, *The End of Imagination* (Haymarket Books, 2016), 335.
8. Eno quoted in Cambridge Audio, "Earth Day 2023: In Conversation with Stuart George and Brian Eno," youtube.com, April 23, 2023.
9. The organization here quoted is called the China Labor Bulletin. The information can be found on the Laudes Foundation website, laudesfoundation.org. See also Eno on the *Sounds Like a Plan* podcast.
10. Quotes and assessment are from James Balmont, "Why Korea Is Hot for Trot, the Cheesiest Pop Imaginable," *Guardian*, June 8, 2022; Doheon Kim, "250 and the Hunt for Ppong in Korea," *Weverse Magazine*, June 27, 2022.

12. Low-Carbonism Music

1. Quotes in this and the next paragraph are from Stuart Hall, "Richard Hoggart, *The Uses of Literacy*, and the Cultural Turn," *International Journal of Cultural Studies* 10, no. 1 (2007), 47. The underlying references in this section are to Walter Ong and Marshall McLuhan.
2. On elemental media, see John Durham Peters, *The Marvelous Clouds: Toward a Philosophy of Elemental Media* (University of Chicago Press, 2015); Nicole Starosielski, "The Elements of Media Studies," *Media+Environment* 1, no. 1 (2019).
3. Quotes in this paragraph from Jonathan Sterne, "The Theology of Sound: A Critique of Orality," *Canadian Journal of Communication* 36, no. 2 (2011), 222.
4. Quotes from Pasek's website, annepasek.com; Anne Pasek, "Fixing Carbon: Mediating Matter in a Warming World" (PhD thesis, New York University, 2019), 7–8, 239.
5. See Bina Agarwal and Sunita Narain, *Global Warming in an Unequal World: A Case of Environmental Colonialism* (Centre for Science and Environment, 1991); Malcolm Ferdinand, *Decolonial Ecology: Thinking from the Caribbean World* (Polity, 2022); Mariko Lin Frame, *Ecological Imperialism, Development, and the Capitalist World-System* (Routledge, 2023); Laurie Parsons, *Carbon Colonialism: How Rich Countries Export Climate Breakdown* (Manchester University Press, 2023); Miriam Lang, Mary Ann Manahan, and Breno Bringel, eds., *The Geopolitics of Green Colonialism: Global Justice and Ecosocial Transitions* (Pluto, 2024).
6. Parts of this paragraph paraphrase and adapt W. J. T. Mitchell, *What Do Pictures Want? The Lives and Loves of Images* (University of Chicago Press, 2010), 10. See also Bernadette

Bensaude-Vincent and Sacha Loeve, *Carbon: A Biography* (Polity, 2023).

7. Pasek, "Fixing Carbon," 222.
8. Part of this paragraph is paraphrased and adapted from Michel Callon, "The Embeddedness of Economic Markets in Economics," in Michel Callon, ed., *The Laws of the Markets* (Blackwell, 1998), 50–1. Another part adapts and paraphrases Marilyn Strathern, ed., *Audit Cultures: Anthropological Studies in Accountability, Ethics, and the Academy* (Routledge, 2000), 3.

13. The Missing Link

1. The opening Tickell quote is from her introduction at We Make Tomorrow. Additional quotes are from the Julie's Bicycle website and promotional material received by email. For the industry report mentioned in the preceding paragraph, see Catherine Bottrill et al., *First Step: UK Music Industry Greenhouse Gas Emissions for 2007* (Julie's Bicycle, 2007).
2. Julie's Bicycle, *Culture: The Missing Link to Climate Action* (Julie's Bicycle, 2021), 10.
3. Quote from Darren Henley introducing a 2021 Julie's Bicycle event called "Culture: The Missing Link—A Lens on Policy," youtube.com.
4. Quote from Julie's Bicycle, *Culture*, 10.
5. Quote from the Group of Friends of Culture Based Climate Action, *Emirates Declaration on Cultural-Based Climate Action* (GFCBCA, 2023), 2.
6. Julie's Bicycle, *Culture*, 10.
7. The human animal's existence as a problem to be solved comes from Erich Fromm via Kermit Pattison, *Fossil Men:*

The Quest for the Oldest Skeleton and the Origins of Humankind (HarperCollins, 2020), 30, 434.

8. Quotes in this and the next paragraph are from Julie's Bicycle, *Culture*, 22.
9. Quotes from, respectively, ibid., 2; Julie's Bicycle, "COP26 Call to Action" (2021); and the webpages for Creative Carbon Scotland (which is now Culture for Climate Scotland, cultureforclimate.scot), Green Music Australia (greenmusic.org.au), and Reverb (reverb.org).
10. My summary in this section is compiled from a range of sources. See, for example, C. Wright Mills, *Power, Politics, and People*, ed. Irving Louis Horowitz (Oxford University Press, 1963); Stuart Hall, *Cultural Studies 1983: A Theoretical History*, ed. Jennifer Daryl Slack and Lawrence Grossberg (Duke University Press, 2016 [1983]); Colin Barker, ed., *Revolutionary Rehearsals* (Haymarket, 2002 [1987]); Cary Nelson and Lawrence Grossberg, eds., *Marxism and the Interpretation of Culture* (Macmillan, 1988); Aijaz Ahmad, *In Theory: Classes, Nations, Literatures* (Verso, 1992); Ellen Meiksins Wood, "A Chronology of the New Left and Its Successors, or: Who's Old-Fashioned Now?," *Socialist Register* 31 (1995); Adolph Reed Jr., *Class Notes: Posing as Politics and Other Thoughts on the American Scene* (New Press, 2000); Kristin Ross, *May '68 and Its Afterlives* (University of Chicago Press, 2002); Michael Denning, *Culture in the Age of Three Worlds* (Verso, 2004); Eric Drott, *Music and the Elusive Revolution: Cultural Politics and Political Culture in France, 1968–1981* (University of California Press, 2011); Gavin Walker, ed., *The Red Years: Theory, Politics, and Aesthetics in the Japanese '68* (Verso, 2020); Stuart Hall, *Selected Writings on Marxism*, ed. Gregor McLennan (Duke University Press, 2021); Ellen Schrecker, *The Lost Promise:*

American Universities in the 1960s (University of Chicago Press, 2021); Vivek Chibber, *The Class Matrix: Social Theory After the Cultural Turn* (Harvard University Press, 2022); Vincent Bevins, *If We Burn: The Mass Protest Decade and the Missing Revolution* (Wildfire, 2023); Melinda Cooper, *Counterrevolution: Extravagance and Austerity in Public Finance* (Zone Books, 2024). For a related discussion of constructionism, see Bruce Robbins, "Belittled Magazine," *Baffler* 81 (2025).

11. Quotes from Francis Mulhern, *Culture/Metaculture* (Routledge, 2000), 137; Terry Eagleton, *After Theory* (Basic Books, 2003), 3. See also Ellen Meiksins Wood, "Marxism Without Class Struggle?," *Socialist Register* 20 (1983); Meaghan Morris, "Banality in Cultural Studies," *Discourse* 10, no. 2 (1988); Angela McRobbie, "Post-Marxism and Cultural Studies," in Lawrence Grossberg, Cary Nelson, and Paula Treichler, eds., *Cultural Studies* (Routledge, 1992).
12. Quote from Adolph Reed Jr., "*Django Unchained*, or, The Help: How 'Cultural Politics' Is Worse Than No Politics at All," *nonsite.org* 9 (2013).
13. For additional background on millennials and neoliberal capitalism, from whom I paraphrase some of my framing here, see Malcolm Harris, *Kids These Days: Human Capital and the Making of Millennials* (Little, Brown, and Company, 2017).
14. Quote from Paul Corrigan and Simon Frith, "The Politics of Youth Culture," in Stuart Hall and Tony Jefferson, eds., *Resistance Through Rituals: Youth Subcltures in Post-War Britain* (Hutchinson, 1976 [1975]), 237, 238. See also Angela McRobbie, *Feminism and Youth Culture* (Macmillan, 1991); Serge Denisoff and Richard Peterson, eds., *The Sounds of Social Change: Studies in Popular Culture* (Rand McNally, 1972).

15. See Sut Jhally, "Stuart Hall: The Last Interview," *Cultural Studies* 30, no. 2 (2016); Lawrence Grossberg, *We All Want to Change the World: The Paradox of the US Left—A Polemic* (Creative Commons, 2015); Simon Frith, "The Remembrance of Things Past: Marxism and the Study of Popular Music," *Twentieth-Century Music* 16, no. 1 (2019); Chibber, *Class Matrix*; Bevins, *If We Burn*; Arthur Borriello and Anton Jäger, *The Populist Moment: The Left After the Great Recession* (Verso, 2023).

14. We Make Tomorrow

1. The ideas in this paragraph are paraphrased and quoted from Tore Størvold, *Dissonant Landscapes: Music, Nature, and the Performance of Iceland* (Wesleyan University Press, 2023), 4, 138–9. Although similar ideas are widely expressed throughout the music world, there is no finer exponent than Størvold.
2. Melody Jue and Rafico Ruiz, "Time Is Melting: Glaciers and the Amplification of Climate Change," *Resilience* 7, nos. 2–3 (2020), 178.
3. Quotes from Kathryn Yusoff and Jennifer Gabrys, "Climate Change and the Imagination," *WIREs Climate Change* 2, no. 4 (2011), 516, 519. The host at We Make Tomorrow mentioned here was dramaturg Anthony Simpson-Pike.
4. Ralph Locke, *Music, Musicians, and the Saint-Simonians* (University of Chicago Press, 1986), 18.
5. Caroline Levine, *The Activist Humanist: Form and Method in the Climate Crisis* (Princeton University Press, 2023). For some musical background, see Nicholas Tochka, *Rocking in the Free World: Popular Music and the Politics of Freedom in Postwar America* (Oxford University Press, 2023).

6. Quote from Levine, *Activist Humanist*, 10.

15. A Behavioral Change Is Gonna Come

1. Helen Prior, "How Can Music Help Us to Address the Climate Crisis?," *Music and Science* 5 (2022), 1, 2, 13. See also Lindsay Fleming and Daniel Levitin, "Waking Up to No Sound: Music Psychology and Climate Action," *Impact: The Journal of the Center for Interdisciplinary Teaching and Learning* 13, no. 1 (2024).
2. Michael Ro et al., "Making Cool Choices for Sustainability: Testing the Effectiveness of a Game-Based Approach to Promoting Pro-Environmental Behaviors," *Journal of Environmental Psychology* 53 (2017).
3. Kristin Hurst and Nicole Sintov, "Guilt Consistently Motivates Pro-Environmental Outcomes While Pride Depends on Context," *Journal of Environmental Psychology* 80 (2022). For the quote in the next paragraph, see Rory Mulcahy, Rebekah Russell-Bennett, and Dawn Iacobucci, "Designing Gamified Apps for Sustainable Consumption," *Journal of Business Research* 106 (2020).
4. For summaries and critiques of gamification and environmentalism, see Paul Dourish, "Implications for Design," in Rebecca Grinter et al., eds., *CHI '06: Proceedings of the SGCHI Conference on Human Factors in Computing Systems* (Association for Computing Machinery, 2006); Jon Froehlich, "Gamifying Green: Gamification and Environmental Sustainability," in Steffen Walz and Sebastian Deterding, eds., *The Gameful World: Approaches, Issues, Applications* (MIT Press, 2015); Georgina Gullén et al., "Gamified Apps for Sustainable Consumption: A Systemic

Review," in Mila Bujić, Jonna Koivisto, and Joho Hamari, eds., *GamiFIN Conference 2022* (CEUR-WS, 2022).

5. See, for example, Evgeny Morozov, *To Save Everything, Click Here: The Folly of Technological Solutionism* (Public Affairs, 2013), 296.
6. See George Graham, "Behaviorism," in Edward Zalta and Uri Nodelman, eds., *The Stanford Encyclopedia of Philosophy* (Stanford University, 2023). See also B. F. Skinner, *Beyond Freedom and Dignity* (Penguin, 1971).

16. No Place

1. Quotes from Tore Størvold, "Confronting Climate Change in Popular Music Texts: Nostalgia, Apocalypse, Utopia," in Mike Dines, Gareth Dylan Smith, and Shara Rambarran, eds., *The Intellect Handbook of Popular Music Methodologies* (Intellect, 2025), 602, 601. See also Josh Kun, *Audiotopia: Music, Race, and America* (University of California Press, 2005).
2. For a few discussions that take seriously both the potentials and limitations of various forms of posthumanism (as well as related ideas like the nonhuman, new materialism, and the ontological turn), see Alexander Weheliye, "Feenin: Posthuman Voices in Contemporary Black Popular Music," *Social Text* 20, no. 2 (2002); Kate Soper, "The Humanism in Posthumanism," *Comparative Critical Studies* 9, no. 3 (2012); Lucas Bessire and David Bond, "Ontological Anthropology and the Deferral of Critique," *American Ethnologist* 41, no. 3 (2014); David Alderson and Robert Spencer, eds., *For Humanism: Explorations in Theory and Politics* (Pluto, 2017); Zakiyyah Iman Jackson, *Becoming Human: Matter*

and Meaning in an Antiblack World (New York University Press, 2020); Marek Susdorf, "Musicolonialism in Suriname: Sonic Contributions to the Construction of the Category of the Human and Its Others" (PhD thesis, University of Oslo, 2021); Peter Osborne, "The Planet as Political Subject?," *New Left Review* 145 (2024); Paul Rekret, *Take This Hammer: Work, Song, Crisis* (Goldsmiths University Press, 2024); Alyssa Battistoni, *Free Gifts: Capitalism and the Politics of Nature* (Princeton University Press, 2025).

3. William Cronon, "The Trouble with Wilderness; or, Getting Back to the Wrong Nature," in William Cronon, ed., *Uncommon Ground: Rethinking the Human Place in Nature* (W. W. Norton, 1995), 69.
4. Quotes from Quinn Slobodian, *Crack-Up Capitalism: Market Radicals and the Dream of a World Without Democracy* (Allen Lane, 2023), 235.
5. Donna Lee King, "*Captain Planet and the Planeteers*: Kids, Environmental Crisis, and Competing Narratives of the New World Order," *Sociological Quarterly* 35, no. 1 (1994), 104, 117.
6. For background, see Nico Baumbach, Damon Young, and Genevieve Yue, "For a Political Critique of Culture," *Social Text* 34, no. 2 (2016); Patricia Stuelke, *The Ruse of Repair: US Neoliberal Empire and the Turn from Critique* (Duke University Press, 2021).
7. Herman Schwendinger and Julia Schwendinger, *The Sociologists of the Chair: A Radical Analysis of the Formative Years of North American Sociology, 1883–1922* (Basic Books, 1974), xvii.
8. Walter Benn Michaels, *The Trouble with Diversity: How We Learned to Love Identity and Ignore Inequality* (Metropolitan Books, 2006), 199–200.
9. On "ecomusicology," the label that much environmental music research applies to itself, see Alexander Rehding,

"Eco-Musicology," *Journal of the Royal Musical Association* 127, no. 2 (2002); Mark Pedelty et al., "Ecomusicology: Tributaries and Distributaries of an Integrative Field," *Music Research Annual* 3 (2022).

17. The Future, Wouldn't That Be Nice?

1. I am paraphrasing Tom Western, "Covered Mouths Still Have Voices," *Journal of Sonic Studies* 24 (2023), 12. See also Herbert Marcuse, *The Aesthetic Dimension: Toward a Critique of Marxist Aesthetics* (Beacon Press, 1978); Ralph Locke, *Music, Musicians, and the Saint-Simonians* (University of Chicago Press, 1986); Catherine Pickstock, "Music: Soul, City, and Cosmos after Augustine," in John Milbank, Catherine Pickstock, and Graham Ward, eds., *Radical Orthodoxy: A New Theology* (Routledge, 1999); Jill Dolan, *Utopia in Performance: Finding Hope at the Theater* (University of Michigan Press, 2005); Sophie Bourgault, "Music and Pedagogy in the Platonic City," *Journal of Aesthetic Education* 46, no. 1 (2012); Stefano Harney and Fred Moten, *The Undercommons: Fugitive Planning and Black Study* (Autonomedia / Minor Compositions, 2013); Michael Denning, *Noise Uprising: The Audiopolitics of a World Musical Revolution* (Verso, 2015); Xiaofei Tian, "The Emperor's New Music: The High Stakes of Popular Music in Ancient China," *Lapham's Quarterly* 10, no. 4 (2017); Martin Stokes, "The Musical Citizen," *Ethnomusicology Journal* 1, no. 2 (2018); Luis Manuel García-Mispireta, *Together, Somehow: Music, Affect, and Intimacy on the Dancefloor* (Duke University Press, 2023); Jasmine Erdener, "Prefigurative Politics at Bread and Puppet Theater," *Cultural Politics* 20, no. 1 (2024).

2. Quotes from Jacques Attali, *Noise: The Political Economy of Music* (University of Minnesota Press, 1985 [1977]), 11; Eric Drott, "Rereading Jacques Attali's *Bruits*," *Critical Inquiry* 41, no. 4 (2015), 754.
3. Christopher Small, *Music, Society, Education: A Radical Examination of the Prophetic Function of Music in Western, Eastern, and African Cultures with Its Impact on Society and Its Use in Education* (Calder, 1977), 2, 227. Note that Small claims not to have agreed to this subtitle. It was apparently added by the publisher without his consent and removed from later editions. Still, his arguments do warrant it. See Mary Cohen, "Christopher Small's Concept of Musicking" (PhD thesis, University of Kansas, 2007), 80.
4. For background, see Matt Stahl, *Unfree Masters: Recording Artists and the Politics of Work* (Duke University Press, 2013); Georgina Born, Eric Lewis, and Will Straw, eds., *Improvisation and Social Aesthetics* (Duke University Press, 2017); Dan DiPiero, *Contingent Encounters: Improvisation in Music and Everyday Life* (University of Michigan Press, 2022); Robert Adlington, *Musical Models of Democracy* (Oxford University Press, 2023).
5. Quote from David Graeber, "At Long Last," in Terence Turner, ed., *The Fire of the Jaguar* (HAU Books, 2017), xxxiv.
6. I am paraphrasing and adapting part of this paragraph from David Graeber, *The Utopia of Rules: On Technology, Stupidity, and the Secret Joys of Bureaucracy* (Melville House, 2015), 89.
7. For background, see Wini Breines, *Community and Organization in the New Left, 1962–1968: The Great Refusal* (Rutgers University Press, 1989); Luke Yates, "Rethinking Prefiguration: Alternatives, Micropolitics, and Goals in Social Movements," *Social Movement Studies* 14, no. 1 (2015); Lara Monticelli, ed., *The Future Is Now: An Introduction to*

Prefigurative Politics (Bristol University Press, 2022); Vincent Bevins, *If We Burn: The Mass Protest Decade and the Missing Revolution* (Wildfire, 2023).

8. The second sentence in this paragraph contains a nod to Jayne Cortez, *Somewhere in Advance of Nowhere* (High Risk Books, 1996). See also Cedric Robinson, *Black Marxism: The Making of the Black Radical Tradition* (University of North Carolina Press, 2000 [1983]; Robin D. G. Kelley, *Freedom Dreams: The Black Radical Imagination* (Beacon Press, 2002); Ingrid Monson, *Freedom Sounds: Civil Rights Call Out to Jazz and Africa* (Oxford University Press, 2007); William Robinson, Salvador Rangel, and Hilbourne Watson, "The Cult of Robinson's *Black Marxism*: A Proletarian Critique," *Philosophical Salon*, October 3, 2022.
9. Kurt Vonnegut spoke of the arts like this in a 1969 speech, in Robert Tally Jr.'s *Kurt Vonnegut and the American Novel* (Bloomsbury, 2011).
10. Quote from Georgina Born, ed., *Music and Digital Media: A Planetary Anthropology* (UCL Press, 2022), 1–2.
11. Quotes from *The Digital Dilemma: Intellectual Property in the Information Age* (National Academy Press, 2000), 78, 79.
12. Quotes from Margaritis Schinas, foreword to William Hammonds, *Culture and Democracy, the Evidence: How Citizens' Participation in Cultural Activities Enhances Civic Engagement, Democracy, and Social Cohesion: Lessons from International Research* (European Commission, 2023), 1.
13. Felicity Laurence quoting the educator in Dave Laing, "Christopher Small Obituary," *Guardian*, September 19, 2011.
14. Quote from Christopher Small, *Musicking: The Meanings of Performing and Listening* (Wesleyan University Press, 1998), 46.
15. Quotes from Timothy Mitchell, "Infrastructures Work on Time," *e-flux*, January 2020, 3; Timothy Mitchell, "Oil,

Climate Crisis, and the Alibi of Growth," lecture, University of Oslo, October 22, 2024. See also Georgina Born, "Future Making: Corporate Performativity and the Temporal Politics of Markets," in David Held and Henrietta Moore, eds., *Cultural Politics in a Global Age: Uncertainty, Solidarity, and Innovation* (Oneworld, 2007); Jenny Andersson, *The Future of the World: Futurology, Futurists, and the Struggle for the Post–Cold War Imagination* (Oxford University Press, 2018); Devon Powers, *On Trend: The Business of Forecasting the Future* (University of Illinois Press, 2019); Liliana Doganova, *Discounting the Future: The Ascendancy of a Political Technology* (Zone Books, 2024).

18. One Thing to Another

1. Léonie Adams et al., *Culture and the Crisis: An Open Letter to the Writers, Artists, Teachers, Physicians, Engineers, Scientists, and Other Professional Workers of America* (Workers Library Publishers, 1932), 3, 30, 29.
2. Quotes from Vincent Bevins, *If We Burn: The Mass Protest Decade and the Missing Revolution* (Wildfire, 2023), 3–4, 3.
3. Adolph Reed Jr., *Class Notes: Posing as Politics and Other Thoughts on the American Scene* (New Press, 2000), xxvii–xxviii.
4. Quote from Marta Harnecker, *Rebuilding the Left* (Zed Books, 2017), 4.
5. Quote from ibid.
6. Quote from Rohini Hensman, *Workers, Unions, and Global Capitalism: Lessons from India* (Columbia University Press, 2011), 318.
7. Michael Denning, "Representing Global Labor," *Social Text* 25, no. 3 (2007), 144. In the next paragraph the quotes are

from Hensman, *Workers*, 318. Regarding the final paragraphs of this section, see Fredric Jameson, *Postmodernism, or The Cultural Logic of Late Capitalism* (Duke University Press, 1991), 209; Michael Denning, *The Cultural Front: The Laboring of American Culture in the Twentieth Century* (Verso, 2010 [1997]), 118; Charlie Post, "The Popular Front Didn't Work," *Jacobin*, October 17, 2017; Shannan Clark, *The Making of the American Creative Class: New York's Culture Workers and Twentieth-Century Consumer Capitalism* (Oxford University Press, 2021); Steve Fraser, "The New Popular Front," *Catalyst* 9, nos. 2–3 (2025).

8. The insight I am paraphrasing and adapting in part of this paragraph, and partly turning around, is from Michael Denning, *Culture in the Age of Three Worlds* (Verso, 2004), 233–4.

Postscript

1. On climate targets, see Andreas Malm and Wim Carton, *Overshoot: How the World Surrendered to Climate Breakdown* (Verso, 2024). With enemy ecology, I'm invoking Sophie Lewis, *Enemy Feminisms: TERFs, Policewomen, and Girlbosses Against Liberation* (Haymarket, 2025).
2. Quotes in this section are from the ClimateEQ website (climate.eq.co.uk), the FEAT blog (feat.ltd/blog), the Good Neighbor website (goodneighbormusic.com), and personal communications.
3. Quotes and paraphrases in this section are from the *Get Wrecked* press release, available on the Pirates Press Records, website (piratespressrecords.com), as well as Khagan Aslanov, "The Vicious Cycles: Born to *Get Wrecked*," *RANGE*, 2025.

4. Lyrics from the Vicious Cycles, "Hold On Tight," *Get Wrecked* (Vicious Cycles MC / Pirates Press Records, 2025). Used with permission.
5. The reference to fun and struggle is adapted from Simon Frith, *Sound Effects: Youth, Leisure, and the Politics of Rock 'n' Roll* (Pantheon, 1981), 274.

Acknowledgments

1. Göran Therborn, *Science, Class, and Society: On the Formation of Sociology and Historical Materialism* (Verso, 1980 [1976]), 415.
2. Michael Burawoy, "On Desmond: The Limits of Spontaneous Sociology," *Theory and Society* 46, no. 4 (2017), 281–2. See also Howard Becker, "Whose Side Are We On?," *Social Problems* 14, no. 3 (1967); Alvin Gouldner, "The Sociologist as Partisan," *American Sociologist* 3, no. 2 (1968); Pierre Bourdieu, *Pascalian Meditations* (Polity, 2000); Anand Pandian, *A Possible Anthropology: Methods for Uneasy Times* (Duke University Press, 2019); Olúfẹ́mi O. Táíwò, "Being in the Room Privilege: Elite Capture and Epistemic Difference," *Philosopher* 8, no. 4 (2020).
3. Some of my framing here is inspired by Anwar Shaikh, *Capitalism: Competition, Conflict, Crises* (Oxford University Press, 2016), 7.

Index